CAMBRIDGE LIBRARY COLLECTION

Books of enduring scholarly value

Mathematical Sciences

From its pre-historic roots in simple counting to the algorithms powering modern desktop computers, from the genius of Archimedes to the genius of Einstein, advances in mathematical understanding and numerical techniques have been directly responsible for creating the modern world as we know it. This series will provide a library of the most influential publications and writers on mathematics in its broadest sense. As such, it will show not only the deep roots from which modern science and technology have grown, but also the astonishing breadth of application of mathematical techniques in the humanities and social sciences, and in everyday life.

Principles of Geometry

Henry Frederick Baker (1866–1956) was a renowned British mathematician specialising in algebraic geometry. He was elected a Fellow of the Royal Society in 1898 and appointed the Lowndean Professor of Astronomy and Geometry in the University of Cambridge in 1914. First published between 1922 and 1925, the six-volume *Principles of Geometry* was a synthesis of Baker's lecture series on geometry and was the first British work on geometry to use axiomatic methods without the use of co-ordinates. The first four volumes describe the projective geometry of space of between two and five dimensions, with the last two volumes reflecting Baker's later research interests in the birational theory of surfaces. The work as a whole provides a detailed insight into the geometry which was developing at the time of publication. This, the second volume, describes the principal configurations of space of two dimensions.

Cambridge University Press has long been a pioneer in the reissuing of out-of-print titles from its own backlist, producing digital reprints of books that are still sought after by scholars and students but could not be reprinted economically using traditional technology. The Cambridge Library Collection extends this activity to a wider range of books which are still of importance to researchers and professionals, either for the source material they contain, or as landmarks in the history of their academic discipline.

Drawing from the world-renowned collections in the Cambridge University Library, and guided by the advice of experts in each subject area, Cambridge University Press is using state-of-the-art scanning machines in its own Printing House to capture the content of each book selected for inclusion. The files are processed to give a consistently clear, crisp image, and the books finished to the high quality standard for which the Press is recognised around the world. The latest print-on-demand technology ensures that the books will remain available indefinitely, and that orders for single or multiple copies can quickly be supplied.

The Cambridge Library Collection will bring back to life books of enduring scholarly value (including out-of-copyright works originally issued by other publishers) across a wide range of disciplines in the humanities and social sciences and in science and technology.

Principles
of Geometry

VOLUME 2:
PLANE GEOMETRY

H.F. BAKER

CAMBRIDGE UNIVERSITY PRESS

Cambridge, New York, Melbourne, Madrid, Cape Town, Singapore,
São Paolo, Delhi, Dubai, Tokyo, Mexico City

Published in the United States of America by Cambridge University Press, New York

www.cambridge.org
Information on this title: www.cambridge.org/9781108017787

© in this compilation Cambridge University Press 2010

This edition first published 1922
This digitally printed version 2010

ISBN 978-1-108-01778-7 Paperback

PRINCIPLES OF GEOMETRY

CAMBRIDGE UNIVERSITY PRESS
C. F. CLAY, Manager
LONDON : FETTER LANE, E.C. 4

LONDON : H. K. LEWIS AND CO., Ltd.,
136, Gower Street, W.C. 1
NEW YORK : THE MACMILLAN CO.
BOMBAY ⎫
CALCUTTA ⎬ MACMILLAN AND CO., Ltd.
MADRAS ⎭
TORONTO : THE MACMILLAN CO. OF
CANADA, Ltd.
TOKYO : MARUZEN-KABUSHIKI-KAISHA

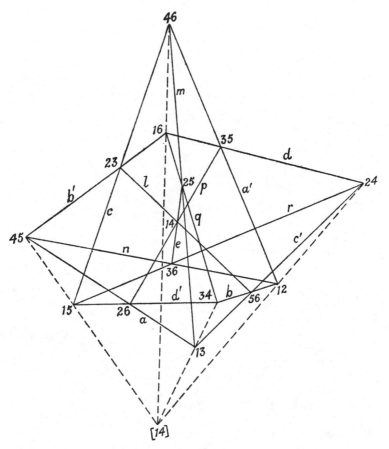

HEXAGRAMMUM MYSTICUM
(see p. 219)

PRINCIPLES OF GEOMETRY

BY

H. F. BAKER, Sc.D., F.R.S.,

LOWNDEAN PROFESSOR OF ASTRONOMY AND GEOMETRY, AND FELLOW OF
ST JOHN'S COLLEGE, IN THE UNIVERSITY OF CAMBRIDGE

VOLUME II

PLANE GEOMETRY

CONICS, CIRCLES, NON-EUCLIDEAN GEOMETRY

In minimis maxima

CAMBRIDGE

AT THE UNIVERSITY PRESS

1922

PRINTED IN GREAT BRITAIN

PREFACE

THE present volume has in effect two aims: In the first place, in pursuance of the general purpose of the book, it seeks to put the reader in touch with the main preliminary theorems of plane geometry. Chapter I is devoted to a deduction, with synthetic methods, of the fundamental properties of conic sections; it is an introduction to what is usually called Projective Geometry, in the plane, in which, however, the notions of distance and congruence are not assumed. Chapter II, also without help of these notions, develops results that arise by considering conics in relation to two Absolute points, including, for instance, the properties of circles, and of confocal conics; the matter here contained is usually found in sequels to Euclid, books on Pure Geometry, and books on Geometrical Conics. Chapter III is designed to explain the application of the algebraic symbols to plane geometry; it contains methods and formulae found in works on Analytical Geometry of the Plane. Chapter IV is a brief consideration of some logical questions, and marks the recognition of a limitation in the symbols employed; it deals with the sense in which the words real and imaginary are used, and calls attention to the elements of Analysis assumed in the following chapter. Chapter V deals with the theory of measurement, of length and angle, with the help of an Absolute conic, shewing how the so-called non-Euclidean geometries may be regarded as included in our general formulation. It considers the metrical plane also as deduced from the geometry of a quadric surface, incidentally dealing with the fundamental properties of this surface and, in particular, with Spherical Trigonometry. As a corollary from this point of view, Riemann's space of constant curvature is seen not to require the assumption of absolute coordinates; and further, that form of the hyperbolic geometry in which lines are replaced by circles cutting a fixed circle at right angles (which, for instance, was an inspiration to Poincaré in his development of the theory of automorphic functions) is seen to arise naturally. Notes I and II deal with the theorems of incidence which were developed very gradually for the complete Pascal figure and appeared very intricate; from the point of view here explained they are natural, if particular, properties of a figure which arises otherwise, and will much concern us in a later volume. Note III gives some indications of the literature of non-Euclidean geometry. Note IV contains remarks and corrections for Volume I, for many of which I am indebted to friends. There is also an Index; but it is possible that the extensive Table of Contents may be more useful. No attempt is made to give a general Bibliography for the contents of the volume.

It will be seen that the volume deals with a wide range of theory; in other conditions than the present, a less condensed treatment might have been desirable. The order in which the ideas are taken has been chosen largely in view of the second aim of the volume; it will not be difficult, with the help of the Table of Contents, for the reader to modify this order. It is believed, however, that a large amount of the time usually spent, at present, in learning geometry, could be saved by following, from the beginning, after an extensive study of diagrams and models, the order of development here adopted; and such a plan would make much less demand upon the memory.

But the second aim of the volume may, I hope, appeal to attentive readers. It is an attempt, tempered indeed by practical considerations, to test the application in detail of the logical principles explained in Volume I. It seeks to bring to light the assumptions which underlie an extensive literature in which coordinates are freely used without attempt at justification. It suggests the question whether, in the case of distance, as in many other cases, we may not have derived from familiarity with physical experiences, a confidence which a more careful scrutiny can only regard as an illusion. When this view, which seems sure, shall win acceptance, the change in scientific thought will be rapid and momentous. As the first step in this sense was made in the development of the theory of our geometrical conceptions, it is proper that the matter should be dealt with here. It will be of importance if the reader come to see how deep lying are the questions involved in the use of coordinates, and the assumption of distance as a fundamental idea.

As in the case of the first volume, I desire to express my thanks to the Staff of the University Press for their care and courtesy, and to Mr J. B. Peace, M.A., for the great trouble he has taken with the numerous diagrams.

H. F. B.

2 *September* 1922

TABLE OF CONTENTS

PRELIMINARY

CHAPTER I.   GENERAL PROPERTIES OF CONICS

CHAPTER II. PROPERTIES RELATIVE TO TWO POINTS OF REFERENCE

CHAPTER III. THE EQUATION OF A LINE, AND OF A CONIC

Contents

CHAPTER IV. RESTRICTION OF THE ALGEBRAIC SYMBOLS. THE DISTINCTION OF REAL AND IMAGINARY ELEMENTS

CHAPTER V. PROPERTIES RELATIVE TO AN ABSOLUTE CONIC. THE NOTION OF DISTANCE. NON-EUCLIDEAN GEOMETRY

PAGES

NOTE I. ON CERTAIN ELEMENTARY CONFIGURATIONS, AND
ON THE COMPLETE FIGURE FOR PAPPUS' THEOREM

PRELIMINARY

THERE are several matters, readily understood by the reader of Volume I, in regard to which we have not there entered into the detail which may be desirable for the purposes of the present volume.

Related ranges on the same line. With the purpose of avoiding the use of points whose existence could only be assumed after the consideration of the so-called imaginary points, we have (Vol. I, pp. 18, 25) defined two ranges on the same line as being *related* when, one of them is in perspective with a range on a second line which is related to the other range of the first line. From this definition we have shewn (Vol. I, p. 160) that in the abstract geometry two such related ranges on the same line have two corresponding points in common, though these may coincide. Assuming this, we may now formally prove that two such ranges also satisfy the general definition, namely that they are both in perspective with the same other range on another line, from different centres.

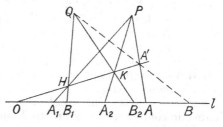

Let the ranges (a), (b), on the same line, l, be such that (a) is in perspective with a range (c), while (c) is related to (b). Let O be a point of the line l which corresponds to itself whether regarded as belonging to the range (a) or to the range (b); let A_1, A_2, A be other points of the range (a), respectively corresponding to the points B_1, B_2, B of the range (b). Let H, K be any two points in line with O; let A_1H, A_2K meet in P, and B_1H, B_2K meet in Q, and let PA meet the line OHK in A'. Then the range O, H, K, A' is in fact related to O, B_1, B_2, B. For the former is in perspective, from P, with the range O, A_1, A_2, A; this is, by hypothesis, in perspective with a range (c), which is itself related to O, B_1, B_2, B; so that the result follows from Vol. I, pp. 22–24. Thence, as the ranges, O, H, K, A' and O, B_1, B_2, B, have the point O in common, they are in perspective (Vol. I, p. 58, Ex. 2 (c)). Thus the line QB passes through A', and the two ranges (a), (b) are in perspective with the same range on the line OHK, respectively from P and Q. This is what we were to prove.

It is clear that the line PQ meets the line l in another common corresponding point of the two ranges (a), (b), which may however coincide with O.

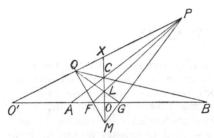

Involution. Considering two related ranges upon the same line, it does not generally follow that, if K be the point of the second range corresponding to a point H of the first range, then to the point K, considered as belonging to the first range, there corresponds the point H, considered as belonging to the second. We proceed to shew, however, that, if this be true for one position, F, of H, and the corresponding position, G, of K, then it is true for every pair of corresponding points H, K.

For, in accordance with the preceding, let the ranges be in perspective, respectively from the points P and Q, with the same range; let X be the point of this range which lies on the line PQ, and O' the common corresponding point of the two ranges arising by perspective from X; let L be the point of this range which gives rise, respectively from P and Q, to the two particular corresponding points F, G of the two ranges, the points P, L, F, and also the points Q, L, G, being in line. Then, by hypothesis, the lines PG, QF meet in a point, M, of the line XL. Let O be the intersection of the line XL with the original line, so that O is also a common corresponding point of the two ranges.

It is clear from the construction that the points O, O' are harmonic conjugates in regard to F and G, and therefore do not coincide with one another; unless, indeed, they both coincide either with F, or with G (I, pp. 14, 119), in which case, as O' is a self-corresponding point, G would coincide with F, which we suppose not to be the case. Thus, when the two common corresponding points of the two given ranges are coincident, the case of two such different reciprocally related points F and G as we are now considering does not arise. Further, the points O', X are harmonic conjugates in regard to P and Q.

Now let A, B be any two other corresponding points, respectively of the two given ranges, arising from the point C of the line XL, by perspective from P and Q, respectively, so that P, C, A, and also Q, C, B are in line. Then, as O', X are harmonic conjugates in regard to P and Q, it follows that O', O are harmonic conjugates in regard to A and B. From this it follows that PB and QA meet on the line XL, and, therefore, that, to the point B, of the first range, corresponds the point A, of the second; as we desired to prove.

Conversely, any two points of the original line which are harmonic conjugates in regard to O' and O are a pair of reciprocally corresponding points of the two ranges. The aggregate of such pairs is called an *involution* of pairs of points; the points O', O are called the *double points* of the involution.

It is clear that if three pairs of points of a line, (A, B), (F, G) and (U, V), be pairs of an involution, then the range A, F, G, U, consisting of two points of one pair, and a point from each of the other two pairs, is related, point to point, to the range B, G, F, V, consisting of the respectively complementary points of the various pairs. For we can define two related ranges by the fact that the points A, F, U, of the one, correspond, respectively, to the points B, G, V, of the other; then, to the point G, of the former, corresponds the point F, of the latter. Conversely, if six points of a line be such that the range A, F, G, U is related, point to point, to the range B, G, F, V, then (A, B), (F, G) and (U, V) are three pairs of an involution. Also, an involution is established when two pairs are given; for, if these be (A, B) and (F, G), we have only to associate, to any point U, a point V for which the range B, G, F, V is related, point to point, to the range A, F, G, V.

Thus, further, a pair of points, O and O', exists, which are harmonic conjugates both in regard to one arbitrary pair of points, A, B, and also in regard to another arbitrary pair, F, G, which lie in the line AB. These points, O, O', are the double points of the involution determined by the pairs A, B and F, G.

From this it follows, also, that if there be two involutions of pairs of points upon the same line, there is a pair of points common to both involutions. This pair consists of the two points which are harmonic conjugates in regard to the double points of the first involution, and also harmonic conjugates in regard to the double points of the second involution.

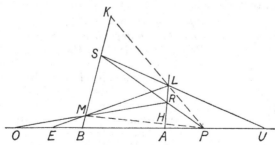

The definition of an involution of pairs of points on a line may be approached differently, in connexion with a figure previously employed (Vol. I, pp. 76, 77).

Let A, B, O, U, E be arbitrary points of a line; in a plane through the line draw two lines AL, BM, met by a line through E respectively in L and M; let UL and BM meet in S, and OM meet AL in R. Let SR meet AB in P.

If, then, PL, PM, respectively, meet BM and AL in K and H, the range O, A, E, P is in perspective, from M, with the range R, A, L, H; this last is in perspective, from P, with the range S, B, K, M, and this, again, in perspective, from L, with the range U, B, P, E. As, then, the ranges O, A, E, P and U, B, P, E are proved to be related, it follows, from what is said above, that the pairs (O, U), (A, B) and (E, P) are in involution.

Thus the three pairs of joins of any four points of a plane (in this case M, R, L, S) meet an arbitrary line, of the plane, in three pairs of points of the same involution. As has been remarked in Vol. I (p. 181), the significance of the fact that five of these points determine the remaining one was noted by the Greeks.

Every result relating to pairs of points of a line corresponds, by the principle of duality, to a result relating to pairs of lines, in a plane, passing through a point. We may therefore consider pairs of lines, in a plane, through a point, forming a pencil in involution; these meet an arbitrary line of the plane, not passing through the centre of the pencil, in pairs of points of a range in involution. In particular, if four arbitrary lines be given in a plane, these, divided into two pairs in each of the three possible ways, determine, by the intersections of lines of a pair, three pairs of points. The lines which join an arbitrary point of the plane to these three pairs of points, are three pairs of lines belonging to the same involution.

Symbolical expression of the preceding results. We have in Vol. I (pp. 140, 154) reached the result that if a range of points represented by symbols $O + xU$, for different values of x, be related to a range of points represented by symbols $O + yU$, for different values of y, then there is a relation of the form

$$ayx + by + cx + d = 0,$$

wherein the symbols a, b, c, d are independent of x and y. We are, throughout, assuming Pappus' theorem, and there is no question of

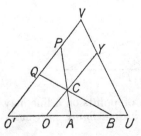

the commutativity of the multiplication of the symbols. It may be interesting, in the first place, to obtain this result by regarding the ranges as being in perspective, from different centres, with the same range on another line.

For this, let the ranges $(A...)$, $(B...)$, on a line $O'U$, be in perspective with a range $(C...)$, on a line OY, respectively, from centres P and Q. Let the line PQ

meet $O'U$ in O'; let V be any point of this line, and let any line from V meet the line of the range $(C...)$ in the point Y, and meet the line $O'U$ in the point U; let the line of the range $(C...)$ meet the line $O'U$ in O. Then, regarding the points O', U, V as fundamental points of the plane, we may suppose, for the symbols of the points concerned, the following expressions

$$P = O' + pV, \quad Q = O' + qV, \quad O = O' + mU, \quad Y = U + V.$$

Thence, if we take for different points C,

$$C = O + cY,$$

with different values of c, that is

$$C = O' + mU + c(U + V), = O' + (m + c)U + cV,$$

we infer, for the points A, B, of the ranges on OU, which arise from C, respectively the symbols

$$(c^{-1} - p^{-1})O' + (c^{-1}m + 1)U, \quad (c^{-1} - q^{-1})O' + (c^{-1}m + 1)U;$$

if we write, then, $A = O' + xU, \quad B = O' + yU$,

this gives

$$\frac{1 + mp^{-1}}{1 - mx^{-1}} = 1 + mc^{-1} = \frac{1 + mq^{-1}}{1 - my^{-1}}$$

and, hence, $(p - q)xy + q(p + m)x - p(q + m)y = 0$.

This is of the form in question. It shews, putting $x = 0$, $y = 0$, that O' is a common corresponding point of the two ranges $(A...)$ and $(B...)$; and, putting $x = m$, $y = m$, that O is the other common corresponding point. It coincides with O' when $m = 0$.

Passing now to the condition for an involution: If, when A takes the present position of B, this takes the present position of A, we must have, beside the previous relation, also the relation

$$(p - q)xy + q(p + m)y - p(q + m)x = 0;$$

from these two relations we obtain, by subtraction,

$$[p(q + m) + q(p + m)][x - y] = 0,$$

and hence, A and B being any particular pair of corresponding points which do not coincide,

$$p(q + m) + q(p + m) = 0.$$

When this is satisfied the relation connecting the values of x and y, in general, is at once seen to reduce to

$$2xy - m(x + y) = 0.$$

We may, however, write

$$(x - y)O = (m - y)(O' + xU) - (m - x)(O' + yU),$$
$$= (m - y)A - (m - x)B;$$

thus the harmonic conjugate of O, in regard to A and B, is (Vol. i, p. 74) of symbol

$$(m - y)A + (m - x)B,$$

or

$$(2m - x - y)O' + [m(x + y) - 2xy]U,$$

and coincides with O' whatever pair of corresponding points A and B may be. Thus O and O' must be different, and the general pair of points of the involution is a pair of points harmonically conjugate in regard to these.

Examples of involution. *Ex.* 1. The condition that the pair of points represented by $O'+xU, O'+yU$, should be harmonic conjugates of one another in regard to the pair of points represented by $O'+aU$, $O'+bU$, being the condition that, for proper symbols p, q,

$$x = (pa+qb)/(p+q), \quad y = (pa-qb)/(p-q),$$

is
$$(x-a)(y-b)+(x-b)(y-a)=0,$$

or
$$xy - \tfrac{1}{2}(x+y)(a+b)+ab=0.$$

This remark, (a), gives the pair of points which are harmonic conjugates of one another in regard to each of two other given pairs of points. In particular, the pair of points harmonically conjugate both in regard to O', U, and in regard to $O'+aU, O'+bU$, is given by $O'+kU, O'-kU$, where $k^2 = ab$. It also gives, (b), the relation for a pair of points belonging to a given involution, and, (c), shews that two given involutions, on the same line, have a common pair of points.

Ex. 2. When $O'+xU, O'+yU$, are, as above, corresponding points of two related ranges, of which O' is a common corresponding point, the relation connecting x and y,

$$(p-q)xy + q(p+m)x - p(q+m)y = 0,$$

is capable of one of the two following forms:

$$\frac{y-m}{x-m} = \sigma \frac{y}{x}, \qquad \frac{1}{y} = \frac{1}{x} + \frac{1}{\lambda},$$

where
$$\sigma = \frac{p(q+m)}{q(p+m)}, \qquad \frac{1}{\lambda} = \frac{1}{p} - \frac{1}{q},$$

according as m is not zero, or m is zero. In general, the relation
$$ayx + by + cx + d = 0,$$

if the equation
$$ax^2 + (b+c)x + d = 0,$$

for which $y = x$, have two different roots α, β, is capable of the form

$$\frac{y-\beta}{x-\beta} = \sigma \frac{y-\alpha}{x-\alpha},$$

where
$$\frac{(\sigma+1)^2}{\sigma} = \frac{(b-c)^2}{ad-bc};$$

but, if $\beta = \alpha$, or $(b+c)^2 = 4ad$, the relation is capable of the form

$$\frac{1}{y-\alpha} = \frac{1}{x-\alpha} + \frac{1}{\lambda},$$

where
$$\lambda = \tfrac{1}{2}(b-c)/a.$$

The general condition for a pair of points, represented by $O' + xU$, $O' + yU$, to belong to an involution is of the form

$$axy + b(x + y) + c = 0.$$

If we choose U for one of the two double points, we have $a = 0$. If then, also, O' be the other double point, we have also $c = 0$. If O' and U be any pair of the involution, we have $b = 0$.

In general, the double points are given by the values of x for which

$$ax^2 + 2bx + c = 0 ;$$

if these coincide ($b^2 = ac$), the involution reduces to one fixed point, for which $x = -b/a$, taken in turn with every other point of the line. This very degenerate case has been excluded from consideration in what has preceded.

Ex. 3. If P, Q and P', Q' be any two pairs of points of a line, and P'' be the harmonic conjugate of P in regard to P' and Q', while Q'' is the harmonic conjugate of Q in regard to P' and Q', the pairs (P, Q), (P', Q'), (P'', Q'') are in involution.

For, by construction, P, P'' and Q, Q'' are pairs of an involution with P', Q' as double points, and the harmonic ranges P, P'', P', Q' and Q'', Q, P', Q' are related; the latter, and, therefore, also the former, is related to the range Q, Q'', Q', P' (Vol. I, p. 25, Ex. 1). This shews that (P, Q), (P'', Q''), (P', Q') are pairs of an involution.

If the points P, Q correspond to the roots of $ax^2 + 2hx + b = 0$, or, say, $F = 0$, being given by $O + x_1 U$, $O + x_2 U$, where x_1, x_2 are the roots of this equation, and, similarly, with the same points of reference O, U, the points P', Q' correspond to the roots of $a'x^2 + 2h'x + b' = 0$, or, say, $F' = 0$, it may be shewn that P'', Q'' correspond, similarly, to the roots of the equation

$$(a'b' - h'^2) F - (ab' + a'b - 2hh') F' = 0.$$

Ex. 4. If A, A'; B, B'; ... be pairs of points of a line, which are in involution, and we take the harmonic conjugate of every one of these in regard to two points, U and V, of the line, so obtaining the pairs P, P'; Q, Q'; ..., then these are also pairs in involution.

For any range P, Q ... is then related to the corresponding range A, B ...; this in turn is related to the range A', B' ...; and this to P', Q' ....

Ex. 5. Given any two pairs of points of a line A, B and A', B', let the pair of points which are harmonic both in regard to A, B and in regard to A', B', be denoted by $(AB, A'B')$; and, therefore, if A'' and B'' be two other points of the line, let $\{(AB, A'B'), A''B''\}$ denote the pair harmonic both in regard to the pair $(AB, A'B')$ and in regard to the pair A'', B''. Shew that the three pairs of points

$$\{(AB, A'B'), A''B''\}, \quad \{(A'B', A''B''), AB\}, \quad \{(A''B'', AB), A'B'\}$$

are in involution. If A, B be given by $f = 0$, where

$$f = ax^2 + 2hxy + by^2,$$

and A', B' be given by $f' = 0$, where $f' = a'x^2 + 2h'xy + b'y^2$, shew that $(AB, A'B')$ are given by

$$\begin{vmatrix} y^2, & -yx, & x^2 \\ a, & h, & b \\ a', & h', & b' \end{vmatrix} = 0,$$

and $\{(AB, A'B'), A''B''\}$ are given by $Hf' - H'f = 0$, where, if $A''B''$ be similarly given by $f'' = 0$, the function H is $a'b'' - 2h'h'' + b'a''$, while $H' = ab'' - 2hh'' + ba''$.

A general abbreviated argument for relating two ranges. Consider two lines, which may coincide, or, more generally, may lie in space of any number of dimensions. Let O, U be two points of the former line, and O', U' be two points of the latter line; if the lines coincide O' and U' will each be in syzygy with O and U, the symbols of O' and U' being each expressible by O and U; if the lines intersect there will be one syzygy connecting the four points. Now, suppose that we have a geometrical construction whereby there is determined a definite point P', of the second line, corresponding to every point P of the first line, and a construction whereby we may conversely pass back from P' to the same point P. The construction may be such as gives, when we start from P, other points $A', B', \ldots$, of the second line, beside P', provided these remain the same for every position of P; and similarly for the construction by which we pass back from P' to P.

For greater definiteness we must also add that the construction must be analogous to those considered in the Third Section of Chap. I of Vol. I (p. 74); it must be such that, if the point P have the symbol $O + xU$, and the point P' the symbol $O' + x'U'$, then x' is determined from x by those laws of combination of the symbols which do not involve the solution of any equation of the second or higher order, the solution of such an equation not being without ambiguity; and x must be determined from x' in a similar way. There is therefore a single algebraic equation, connecting x and x', of the form $\Sigma ax'^m x^n = 0$, containing only a finite number of terms, in which every coefficient a is quite definite, m and n being positive integers. Then, as, to every value of x, there belongs only one value of x', other than those independent of x, the equation may be expected to be such that the highest value of the exponent m is 1; and, similarly, such that the highest value of n is also 1. (Cf. Chap. IV, below.) The relation would then be of the form

$$ax'x + bx' + cx + d = 0.$$

When we have such a determinate construction we shall, some-

times, assume, for the sake of brevity, that it is possible to find a
third range with which both the given ranges are in perspective,
from appropriate centres; so that the given ranges are related.

**Of the distinction between the so-called real and imagi-
nary points.** In the general discussion introductory to Chap. III
of Vol. I (p. 141), and in some other cases, we have spoken of the
distinction between real and imaginary points in a way which, if
definite when we approach the matter from the point of view of the
Real Geometry, is not so clear from the point of view to which we
desire to reach. It might be proper then to enter now into more
detail. But we do in fact regard the distinction as arising in connexion with an arbitrary limitation of the possibilities of the points
which can exist in the space considered; for this reason we postpone
this discussion until (in Chap. IV, below) we definitely agree to make
this limitation. This limitation is represented by a restriction in
the system of symbols appropriate to the geometrical results
obtained. While we wish to leave the logical possibilities as open
as we can, we desire to expound a system of geometry in harmony
with what is commonly accepted. For this purpose we have suggested, in Vol. I, that we may take as a geometrical postulate the
possibility of what we have called Steiner's construction (Vol. I,
pp. 155 ff.). In a similar way we shall in the first three chapters
of the present volume adopt as a postulate the theorem that *two
curves in a plane which are *conics*, in the sense to be immediately
explained, *have four common points, of which two, or more, may
coincide*. It will be seen below (in Chap. IV) how this would be
proved when the limitation referred to is adopted. Cf. pp. 19, 157
below.

CHAPTER I

GENERAL PROPERTIES OF CONICS

Definition of a Conic. Consider two pencils of lines, in the same plane, of centres A and C, and suppose these are related to one another, in the sense explained in Vol. I; to any line, AP, of one pencil, there corresponds, then, a definite line, CP, of the other, and, conversely, to any line, CP, there corresponds one line, AP. The curve which is the locus of the intersection, P, of such corresponding rays AP, CP, is called a conic section, or, briefly, a conic. Conversely, it will be seen that any conic can be so obtained, from any two points, A, C, of itself.

By its definition the curve contains one point, P, beside A, upon any line drawn through A; though, when the line drawn through A is that which corresponds to the line CA drawn through C, the point P coincides with A. Similarly for a line drawn through C. Consider however any line not passing through A nor C. This line contains two points of the conic. For the related pencils, of centres A and C, determine upon this line two related ranges. These have two common corresponding points. If these be P_1 and P_2, the ray AP_1, of the one pencil, corresponds to the ray CP_1, of the other, and P_1 is on the locus; and it is the same for P_2.

There is however one case, which we regard as exceptional, to which reference should be made: It may be that, to the ray AC of the pencil (A), there corresponds the ray CA of the pencil (C). When this is so, the locus consists of a line of the plane, together with the line AC itself. For if P_1 and P_2 be, then, any two points of the locus, not lying on the line AC, and the line P_1P_2 meet the line AC in B, the two related pencils (A), (C) determine, on the line P_1P_2, two related ranges having three common corresponding points, namely P_1, P_2 and B. These two ranges thus coincide, and every point of the line P_1P_2 is a point of the locus. If there were any point, P, of the plane, not lying on the line P_1P_2 or the line AC, which was upon the locus, then the join of P to any point of P_1P_2, would, by what has been said, be a line of which every point was a point of the locus; and in that case every point of the plane would be a point of the locus. The complete locus thus consists of the line P_1P_2, and of the line AC, of which every point evidently satisfies the definition. Conversely, any two lines of the plane may be regarded as constituting a degenerate conic, determined by related pencils whose centres are any two points of one of the lines. Such a degenerate conic may be spoken of as a line pair. In Vol. ɪ (p. 58), we have seen that two related ranges on two intersecting lines, which are such that the intersection of the lines is a common corresponding point, are in perspective. The aggregate of the lines joining corresponding points of the two ranges consists then of a pencil of lines, together with every line through the common point of the ranges. This is the result dually corresponding to that here remarked.

In what follows we shall, in general, suppose that, to the ray, AC, of the pencil (A), there corresponds a ray, CB, of the pencil (C), which does not coincide with CA. To the ray CA, of the pencil (C), will then correspond a ray, AB, of the pencil (A), meeting the former, say, in B. As the rays AC, CB meet in C, and the rays CA, AB meet in A, the locus contains the points A and C.

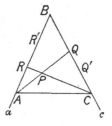

The lines CB, AB, say, respectively, c and a, are determined when the related pencils are given. Conversely, if these lines be given, and, upon them, respectively, the points C and A, and, also, a point, P, of the locus, be given, then the related pencils are determined; for, the three rays AC, AB, AP, of the one, then correspond, respectively, to the three rays CB, CA, CP, of the other. These lines c, a contain no other point of the locus beside, respectively, C and A. For, if AP, CP meet c and a, respectively, in Q and R, there are two related ranges on c and a, in which the points C, B, Q, of the one, correspond, respectively, to the points B, A, R, of the other; if then the point, P, of the locus, lie on c, the point R will be at B, and, therefore, Q, and hence also P, will be at C. Moreover if, for a moment, we limit ourselves to only real points, and take a further point, P', of the locus, for which the corresponding Q, say Q', is separated from B by Q and C, then the corresponding R, say R', will be separated from A by R and B. We therefore speak of the locus as having two *coincident* points at C; similarly, it has two coincident points at A. The lines c and a are called *tangents* of the conic, respectively at C and A. It will appear that, through every point of the conic, a line can be drawn containing only this point of the locus; this line will be called the tangent at the point.

We have remarked that, if the points A, C, and the tangents a, c at these points, and also a point, P, of the locus, be given, the related pencils, and, therefore, also the conic, are determined. More generally, if the points A, C, and three points P_1, P_2, P_3 (of which we suppose at most one, but, in general, none, to lie on the line AC)

be given, the conic is determined; for we have only to make the
rays CP_1, CP_2, CP_3, of the pencil (C), correspond, respectively, to
the rays AP_1, AP_2, AP_3, of the pencil (A), whereby the pencils are
related. Thus a conic can be described through five points of the
plane, of general position; when three of these are in line, the conic
becomes a line pair; when four of them are in line the conic is not
determined.

It has been stated that the points A, C, in the definition of the
conic, may be replaced by any other two points of the locus, say A'
and C'. This is easy to prove. Let O be any
point of the conic, and let any line, drawn
through O, meet the conic again in a point P.
Let the lines AA', CC' meet the line OP,
respectively, in X and Y, and the lines AC',
$A'C$ meet this line, respectively, in D and E.
Then, the pencil, of centre A, formed with
lines joining this point to A', C', O, P, is, by
hypothesis, related to the pencil, of centre C,
formed with lines joining this point to the
same four points. Wherefore, the range X,
D, O, P is related to the range E, Y, O, P,
and, therefore, also to the range Y, E, P, O. Thus (Preliminary,
above, p. 3), the three pairs of points X, Y; D, E; O, P are in
involution. To prove, now, that the conic defined with related
pencils of centres A' and C' to pass through the points A, C, O,
also passes through P, whatever point of the original conic P may
be, we require only to shew that the pencil of lines joining A' to
A, C, O, P is related to the pencil of lines joining C' to these same
points, respectively. This will be so if the range X, E, O, P is
related to the range D, Y, O, P; which will be so if the range
X, E, O, P is related to the range Y, D, P, O. The condition for
this, however, is the same as before, that the pairs of points X, Y;
D, E; O, P should be in involution.

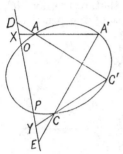

The result is therefore established. Thereby, there is also defined,
at every point of the curve, a line, called the tangent, meeting the
curve only at this point.

**Involution on a line by conics having four common
points.** Incidentally there appears from the preceding proof a
theorem which is of great importance: If four points (A, C, A', C')
be taken on a conic, and two pairs of lines $(AC', A'C$ and $AA', CC')$
of which each pair contains the four points, then an arbitrary line
is met by these line pairs, and by any conic passing through the
four points, in three pairs of points belonging to the same involution. We have seen that a conic can be drawn through five
general points, so that an infinite number of conics can be drawn

through four general points; we have also seen (Preliminary, above, p. 3) that an involution is determined by two of its pairs. Hence we have the result that, the pairs of points in which the conics, drawn through four general points, cut an arbitrary line, are pairs of the same involution. Any one of the three line pairs which can be drawn through these four points is a particular degenerate conic of this description.

MacClaurin's definition of a conic. Pascal's theorem.

It has been shewn in Vol. i (pp. 21 ff.) that two related ranges, on different lines, *a* and *c*, are both in perspective with another range, on a line *o*, say, from proper centres, say *H* and *K*, respectively. The dually corresponding result is that, two related pencils, with different centres, *A* and *C*, must consist of lines joining *A* and *C* to the points, respectively, of two related

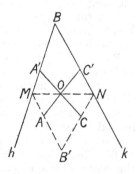

ranges on different lines, say *h* and *k*, these ranges being in perspective with one another from some centre, say *O*.

Thus the definition of a conic which has been given is the same as, the locus of a point, *B'*, which is the intersection of two lines *MB'*, *NB'*, passing, respectively, through two fixed points, *A* and *C*, where *M* and *N* are any two points on two given lines, *h* and *k*, such that *MN* passes through a third fixed point, *O*. Conversely, the ranges (*M*), (*N*) being in perspective, the pencils *A*(*B'*) and *C*(*B'*) are clearly related, so that the present definition leads back to the former.

If *B* be the intersection of the lines *h* and *k*, and *OA*, *OC* meet the lines *k* and *h*, respectively, in *C'* and *A'*, the conic contains the points *B*, *A'*, *C'*. For, first, when *M* and *N* coincide at *B*, the lines *MA*, *NC* intersect in *B*; and, next, when *M* is at *A'*, and, therefore, *N* is at the point where *OC* meets *k*, the line *NC* contains *A'*, which is then the intersection of *NC* and *MA*; similarly, when *N* is at *C'*, so also is *B'*. As was remarked above, the locus contains also the points *A* and *C*. We have thus, at once, a construction, with lines only, for finding any number of points of a conic passing through five arbitrary given points *B*, *A*, *C*, *A'*, *C'*; as was seen, such five points determine the conic. And, from the symmetry which was shewn to hold among the fundamental points, we can now infer that, if *A*, *B'*, *C*, *A'*, *B*, *C'* be any six points of a conic, the three points of intersection of the pairs of lines, *BC'*, *B'C*; *CA'*, *C'A*; *AB'*, *A'B*, are in line. This result is the celebrated one known as Pascal's Theorem. From six points, we can form two sets of three in ten ways, and either of these sets can be arranged, relatively to the other, in six ways; thus, from six given points of a conic can be deduced sixty different lines, each containing three such points of intersection as those here remarked. When the conic consists of two lines, one containing the points A, B, C, the other the points A', B', C', Pascal's Theorem becomes that which, in Vol. I, we have called Pappus' theorem.

The construction at once gives the tangent of the conic at the point A: If the line AC meet the line k in N_0, and the line N_0O meet the line h in T, then AT is the tangent at A. So, if AC meet h in M_0, and M_0O meet k in U, then CU is the tangent at C. The tangent at A thus appears as the position of the line AB' when B' is at A, or, equally, as the ray of the pencil of centre A which corresponds to the ray CA of the pencil of centre C; and similarly for the tangent at C. To find the points, B', of the conic, which lie on an arbitrary line, it is necessary to solve the problem considered in Vol. I, pp. 155 ff., of finding lines through three given points whose intersections lie on three given lines.

Ranges of points on a conic. Since the pencils of four lines which join any four given points of a conic to other points of the conic are all related, the definition of a conic ascribes a definite character to four points of the curve in relation to any other four points: two such sets of four points of the locus may be said to be related to one another when the pencil of four lines, joining the first set of four to any point of the conic, is related to the pencil of four lines joining the second set of four points to any point of the conic. Thus, given any two triads of points of the conic, say A, B, C and A', B', C', we can, to any point, D, of the conic, determine a point, D', of the curve, such that, in the sense explained, the range A', B', C', D', of points of the conic, shall be related to the range A, B, C, D. There exists, therefore, a theory of related ranges of points of the conic precisely like that of related ranges of points of a line.

Involution of pairs of points of a conic. In particular, there is a theory of pairs of points of the conic which are in involution, one form of the condition that three pairs of points of the conic, P, P'; Q, Q'; R, R', be in involution, being, as in the case of points of a line (Preliminary, above, p. 3), that the range P, Q, R, R' should be related to the range P', Q', R', R. This condition, however, is capable of very simple geometrical interpretation: The necessary and sufficient condition that the three pairs of points of the conic, P, P'; Q, Q'; R, R', should be in involution, is that the lines PP', QQ', RR' should meet in a point.

This fact is easy to prove from what we have seen above. Take any three pairs of points of the conic, P, P'; Q, Q' and R, R'; let the line RR' meet the line PP' in U, and meet the line QQ' in V;

further, let the lines $P'Q$ and PQ' meet the line RR' in L and M. By what we have seen, on the line RR' the pairs of points R, R'; U, V and L, M are in involution; therefore, the range U, L, R, R' is related to the range V, M, R', R. The former range, U, L, R, R', is, however, the section, by the line RR', of the pencil of lines joining to P' the points P, Q, R, R' of the conic. The condition that the pairs of points of the conic, P, P'; Q, Q' and R, R', should be in involution, is that the pencil joining P, Q, R, R' to any point of the conic should be related to the pencil joining

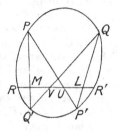

P', Q', R', R to any point of the conic. The lines joining, to P, these last four points, meet the line RR' in, respectively, U, M, R', R. The condition is, then, that the range U, M, R', R should be related to the range U, L, R, R'; this, however, we have remarked, is related to the range V, M, R', R. The condition, therefore, is, that the ranges, U, M, R', R and V, M, R', R, should be related, or that V should coincide with U. This is the condition that the lines, PP', QQ', RR', should meet in one point, as was stated.

Pascal's theorem and related ranges of points upon a conic. From the existence of related ranges upon a conic, we can at once deduce Pascal's theorem. And another proof of the preceding theorem, in regard to involutions of points upon the conic, offers itself immediately.

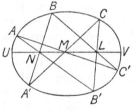

Let A, B, C, A', B', C' be any six points upon the conic; let L be the intersection of the lines $BC', B'C$, and, similarly, M the point $(CA', C'A)$ and N the point $(AB', A'B)$. We can, we have seen, determine two related ranges of points upon the conic by the condition that the points A', B', C', of one of these, correspond, respectively, to the points A, B, C of the other. These ranges will have two common corresponding points; let these be U and V, which, for the present, we suppose not to coincide. Then the pencil of lines joining the points U, V, A, B to any point of the conic is, by hypothesis, related to the pencil joining, respectively, U, V, A', B' to any point of the curve. If the former point be taken to be A', and the latter to be A, and we consider the sections of these pencils by the line UV, so obtaining two sets of four points of which three are the same in both, we infer at once that the point N is on the line UV. Similarly M, and L, lie on this line; thus L, M, N are in line. This is Pascal's theorem.

The condition that the ranges U, V, A, B and U, V, A', B' should be related is the same as that U, V, A, B and V, U, B', A' should be related. This, however, is the same as that the pairs U, V; A, B' and A', B, of points of the conic, should be in involution; for this, then, the lines UV, AB' and $A'B$, must meet in a point. Conversely, Pascal's theorem is equivalent to saying that, if A, B, C, A', B', C' be any six points of the conic, the three involutions determined, respectively, by the two pairs B, C' and B', C; by the two pairs C, A' and C', A; and by the two pairs A, B' and A', B, have a pair common to all three involutions (the pair U, V).

If H, K be any two points of the range $A, B, C, ...$, of points of the conic, and H', K' be the respectively corresponding points of the related range $A', B', C', ...$, the lines $HK', H'K$ can similarly be shewn to meet on the line UV. This line may be called the *axis of relation* of these two ranges, or the axis of homography; cf. Vol. I, p. 53.

Pascal's theorem remains equally true when the points U, V coincide. The line UV is then replaced by the tangent of the conic at U.

Converse of Pascal's theorem. It is easy to prove from the direct theorem that, if A, B, C, A', B', C' be such six points of the plane that the three intersections $(BC', B'C)$; $(CA', C'A)$ and $(AB', A'B)$ are in line, then the six points all lie on a conic.

Introduction of the algebraic symbols. We have thus obtained many of the fundamental theorems for a conic immediately from the definition. It is our purpose to make the geometrical theory complete in itself. But it may be interesting, at the same time, to consider the application of the algebraic symbols.

Returning to the definition, the point, P, of the locus, being the intersection of the corresponding rays AP, CP of two related pencils, of centres A and C, let AP meet in Q the tangent CB at the point C, and CP meet in R the tangent AB at the point A. Then Q, R describe related ranges on the lines CB, AB, of which the points C, B, of the former, correspond, respectively, to the points B, A, of the latter.

If then we represent the point Q by the symbol $C + \theta B$, where θ is an algebraic symbol, we can, absorbing a proper multiplier in the symbol A, represent R by the symbol $B + \theta A$. Thence the point P is given by the symbol

$$P = \theta^2 A + \theta B + C,$$

and all points of the conic are given by this, for different symbols θ. In particular, the points A and C arise, respectively, for $\theta^{-1} = 0$ and $\theta = 0$. We might equally, however, have represented the point Q by $C + m\theta B$, where m is an arbitrary algebraic symbol; to fix the symbol θ for every point of the conic, it is necessary to fix it for one point beside A and C; say, by specifying the point of the curve for which $\theta = 1$.

That a locus, of which every point is given by a symbol such as that here representing P, contains two points of an arbitrary line, is clear, by remarking that the point

$$x_1 A + y_1 B + z_1 C + \lambda \, (x_2 A + y_2 B + z_2 C),$$

of the line joining the given points $x_1 A + y_1 B + z_1 C$, $x_2 A + y_2 B + z_2 C$, will be a point of the conic if, and only if,

$$x_1 + \lambda x_2 = \theta \, (y_1 + \lambda y_2) = \theta^2 \, (z_1 + \lambda z_2) ;$$

by elimination of λ from these two equations we obtain a quadratic equation for θ.

The result that the conics through four points, A, C, A', C', meet an arbitrary line in pairs of points in involution, which was proved in order to shew that the two points A, C, used in the definition of the conic, may be replaced by any other two points A', C', of the conic, may be obtained thus: there is, as was proved, one conic through the four points A, C, A', C' and an arbitrary point, P, of the line. This conic will meet the line in another point, say P'. Thus, to any position of P on the line, there corresponds one and only one position of P' on the line. Conversely, the conic through A, C, A', C' and P' meets the line in P. The determination of P' from P, or of P from P', is clearly by a rational algebraic equation. Thus (see the remarks above, Preliminary, p. 8), (a) the ranges (P), (P'), on the line, are related, (b) when P assumes the position P', then P' assumes the position P. Hence the pairs P, P' belong to an involution.

The existence of related ranges of points on the conic follows from the expression of the points of the conic by means of the parameter θ. In fact, the line joining the arbitrary fixed point, H, of the conic, for which the symbol is $\alpha^2 A + \alpha B + C$, to the point, P, of symbol $\theta^2 A + \theta B + C$, contains the point $(\alpha \theta^2 - \theta \alpha^2) B + (\theta^2 - \alpha^2) C$, that is the point $\alpha \theta B + (\theta + \alpha) C$, of the line CB. Putting $M = B + \alpha^{-1} C$, for the symbol of a definite fixed point on the line CB, the point in question is $\theta M + C$. Similarly the line joining the fixed point, K, of symbol $\beta^2 A + \beta B + C$, to the point P, meets the line CB in the point $\theta N + C$, where $N = B + \beta^{-1} C$, is the symbol of another fixed point of the line CB. The pencil of lines joining H to four points, P, is thus related to the pencil of four lines joining K to these same four points. Further, a range of various positions of points P of the conic, expressed by various values of the parameter θ, is related to a range of positions, Q, expressed by values of ϕ, when there is a condition of the form

$$a\theta\phi + b\theta + c\phi + d = 0,$$

in which a, b, c, d are independent of θ and ϕ. The pairs of corresponding positions, P, Q, then belong to an involution when, in this condition, $c = b$. In other words, points P, Q, of respective parameters θ, ϕ, form a pair of an involution of points on the conic, when θ, ϕ are connected by an equation

$$a\theta\phi + b(\theta + \phi) + d = 0.$$

When this is so, we at once see that

$$(\phi a + b)(\theta^2 A + \theta B + C) - (\theta a + b)(\phi^2 A + \phi B + C)$$

is equal to

$$(\phi - \theta)(dA - bB + aC),$$

so that the line PQ passes through the fixed point $dA - bB + aC$.

But the fact, that the lines drawn through an arbitrary point, O, of the plane, meet the conic in pairs of points, P, Q, of an involution, is clear, also, by remarking, (a), that, when P is given, Q is determined without ambiguity, by the line OP, which meets the conic again in Q; and that P is similarly determined when Q is given, the determination being by a rational algebraic equation connecting the parameters of P and Q; and (b) that when P takes the position Q, then Q takes the position P. The converse result, that, if P, P'; Q, Q' and R, R' be pairs of points of the conic which are in involution, the lines PP', QQ', RR' meet in a point, follows from this if we recall that an involution is determined by two of its pairs.

Pascal's theorem can also be proved at once with the symbols. For let A, B, C, A', B', C' be six points of the conic; and let U, V be the common corresponding points, on the conic, of the two related ranges A, B, C,... and A', B', C',.... In the first instance suppose U and V not to coincide. Let the tangents at U and V meet in W; the points of the conic can then be represented, as above, in terms of a parameter θ, by a formula $\theta^2 U + \theta W + V$. Let $\alpha, \beta, \gamma, \alpha', \beta', \gamma'$ be the parameters for A, B, C, A', B', C' respectively. For corresponding points, with parameters θ, θ', of the two related ranges, there will be a condition of the form

$$p\theta\theta' + \theta' - q\theta + r = 0,$$

where p, q, r are to be determined by the fact that this equation is satisfied by $\theta = \alpha$, $\theta' = \alpha'$; by $\theta = \beta$, $\theta' = \beta'$; and by $\theta = \gamma$, $\theta' = \gamma'$. By hypothesis, however, the equation

$$p\phi^2 + (1 - q)\phi + r = 0,$$

which gives the common corresponding points of the ranges, is satisfied by $\phi = 0$ (at the point V), and by $\phi^{-1} = 0$ (at the point U). Thus $p = 0$ and $r = 0$. Therefore, for a proper value of q, we have

$$\alpha' = q\alpha, \ \beta' = q\beta, \ \gamma' = q\gamma.$$

These relations shew that the lines AB' and $A'B$ meet on the line

UV, namely in the point whose symbol is $q\alpha\beta U - V$. For this is $\alpha\beta'U - V$, or

$$[\alpha - \beta']^{-1}[\beta'(\alpha^2 U + \alpha W + V) - \alpha(\beta'^2 U + \beta'W + V)],$$

and this point, therefore, lies on the line AB'; and it similarly lies on $A'B$. By the same argument the lines BC', $B'C$ meet on the line UV, as do the lines CA', $C'A$. This establishes Pascal's theorem.

If the common corresponding points of the two related ranges, A, B, C,... and A', B', C',..., on the conic, be coincident, say in U, we may take an arbitrary point, V, of the curve, and, as before, represent any point of the curve by a symbol $\theta^2 U + \theta W + V$. The equation, $p\phi^2 + (1 - q)\phi + r = 0$, for the common corresponding points of the two related ranges is, now, to be satisfied by $\phi^{-1} = 0$ (at the point U), *twice over*; thus $p = 0$ and $q = 1$. The relation for the corresponding points of the two ranges is, therefore, $\theta' = \theta - r$, so that $\alpha' = \alpha - r$, $\beta' = \beta - r$, $\gamma' = \gamma - r$. The line AB' now contains the point, of the line UW, given by

$$\alpha^2 U + \alpha W + V - (\beta'^2 U + \beta'W + V);$$

this is the point given by $(\alpha + \beta')U + W$, or $(\alpha + \beta - r)U + W$. By the symmetry of this form, the line $A'B$ also passes through this point. Similarly the intersection of the lines BC', $B'C$, and the intersection of the lines CA', $C'A$, lie on this line UW, which is the tangent of the conic at U.

We have stated, in the Preliminary (above, p. 9), that, in the first three chapters of this volume, we propose to assume that two conics in the same plane have four common points. It may be desirable to remark on the representation of this fact in terms of the symbols. The expression for a point of a conic, $\theta^2 A + \theta B + C$, is in terms of two points A, C, lying on the conic, and the intersection, B, of the tangents of the conic at these two points. If U, V, W be three arbitrary points of the plane, the points A, B, C will be expressible in terms of these by formulae $A = a_1 U + a_2 V + a_3 W$, $B = b_1 U + b_2 V + b_3 W$, $C = c_1 U + c_2 V + c_3 W$ (Vol. I, p. 71). Thus in terms of U, V, W, a point of the conic will have a representation

$$(a_1\theta^2 + b_1\theta + c_1)U + (a_2\theta^2 + b_2\theta + c_2)V + (a_3\theta^2 + b_3\theta + c_3)W.$$

If now there be another conic, the points of this may, similarly, be represented in terms of U, V, W, the expression being of the form

$$(l_1\phi^2 + m_1\phi + n_1)U + (l_2\phi^2 + m_2\phi + n_2)V + (l_3\phi^2 + m_3\phi + n_3)W,$$

where ϕ is the parameter. At a point common to the two conics, the three functions

$$(a_r\theta^2 + b_r\theta + c_r)/(l_r\phi^2 + m_r\phi + n_r),$$

for $r = 1, 2, 3$, must be equal to one another. From the two equations which express this fact we may eliminate ϕ; the result is at once seen to be a quartic equation to be satisfied by θ. The assumption

2—2

to be made is then equivalent to the assumption that this equation
has four roots. From any one of these, the corresponding value of
ϕ can be found without ambiguity.

With these indications, we now leave the use of the symbols.
They will be employed again in Chapter III.

Geometrical theory resumed. Polar lines. We saw that
any line of the plane contains two points of the conic. We now
shew that through any point, O, of the plane, not lying on the curve,
two tangents of the curve pass. Let lines be drawn through O, each
meeting the curve in a pair of points, say P and P'. The pairs P, P'
have been shewn to belong to an involution on the curve; let M
and N be the points of the curve which are the double points of
this involution. The line OM will then meet the curve only at M,
in two coincident points, and will be the tangent of the curve
at M. So the line ON is the tangent at N. And there is no other
tangent passing through O, since the involution has only two double
points.

Moreover, an involution is, by definition, formed by the aggregate
of two related ranges of points which are such that two corresponding points, one from each range, are interchangeable. Thus if,
through the point O, be drawn various lines OPP', OQQ', ORR', ...,
meeting the conic, respectively, in P and P', in Q and Q', in R and
R', ..., the ranges, $P, Q, R, ...$ and $P', Q', R', ...$, are related, and M, N
are the common corresponding points of these two ranges. Thus,
by what has been proved, in particular by Pascal's theorem, the

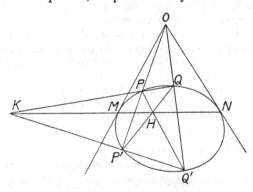

lines PQ', $P'Q$ meet on the line MN, say, in H, as do the lines PQ,
$P'Q'$, say, in K. It follows that the line MN can be constructed
when the four points P, P', Q, Q' are given; it also follows that the
line PP' is met by MN in a point which is the harmonic conjugate
of O in regard to P and P'. The line QQ' is similarly met by MN
in a point which is the harmonic conjugate of O in regard to Q and

Q'; with a similar statement in regard to every one of the lines ORR', ... Thus, if lines be drawn from a point, O, not lying on the conic, and upon each be taken the harmonic conjugate of O in regard to the two points in which this meets the conic, the locus of this fourth harmonic point is a line. This line is called the *polar line*, or simply the *polar*, of O. From this there follows at once a property which is often applied, that, if the polar of one point O pass through a point O', then the polar of O' passes through O. And, therefore, if two points of the line MN be given, then O is determinate, namely, as the intersection of the polar lines of these two points.

When O is on the conic, the harmonic conjugates of this in regard to the pairs P, P' and Q, Q', of which, say, P and Q coincide with O, also coincide with O. In this case we speak of the tangent of the curve at the point O as being the polar of O. Conversely, it may be shewn to follow from the character of the harmonic relation of four points that the polar of O can only pass through O when O is a point of the conic.

We may speak of the point O as the pole of the line MN; and in particular of a point of the conic as the pole of the tangent line of the conic at this point.

Duality in regard to a conic. From what we have just said, a conic, given in a plane, determines a particular duality therein. Any point, P, of the plane, regarded as belonging to one figure, determines a particular line, p', which we may regard as belonging to a derived figure, this line p' being here, the polar of P in regard to the given conic. Any line, p, of the first figure, joining two points, Q and R, of the first figure, gives rise to a point, P', of the derived figure, this being the intersection of the polars, respectively q' and r, of Q and R. To a point H, of the first figure, which coincides with P', corresponds the line, h', of the second figure, which is the polar line of P'; but, as P' lies on the polars of Q and R, these points, Q, R, lie on the polar of P', which is the line h'. Thus, the lines h' and

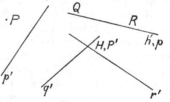

p are the same. The correspondence is, therefore, of an involutory, or reciprocal, character, the line which arises from any point being the same whether the point is regarded as belonging to the first, or to the derived, figure; the point which arises from any line is therefore, also, capable of both derivations. The necessary and sufficient condition for a line to contain the point to which it corresponds is, that the point lie on the conic, and that the line be a tangent of the conic. We may, therefore, speak of the conic as the *curve*, or *envelope, of incidence*, of the two figures.

Dual definition of a conic. When one figure is the dual of another, it follows from the original definition of related ranges, by means of incidences, as developed in Vol. I, and the fact that Pappus' theorem involves the dual proposition which corresponds to it (Vol. I, chap. I, and p. 55), that related ranges give rise to related pencils, and related pencils give rise to related ranges. Hence, as we defined the conic as a locus of points, by the intersection of corresponding rays of two related pencils, so, equally well, we could have defined the aggregate of the tangents of the conic as the joins of corresponding points of two related ranges. In other words, if the tangent at any point, P, of the conic, meet the two tangents at the points A and C, respectively, in T and U, the range of points, T, upon AB, is related to the range of points, U, upon BC, the point U being at B when T is at A, and at C when T is at B.

In fact, the range of points, U, where the tangent, CB, at a fixed point, C, of the curve, is met by the tangent at a variable point P, is related to the pencil of lines joining the points P to any fixed point of the curve. To prove this, if AP meet the tangent CB in Q, the point A being any fixed point of the curve, it is sufficient to prove that the range of points U is related to the range of points Q.

That this is so follows by the abbreviated argument referred to above (Preliminary, p. 8). For when U is given, P is determined, without ambiguity, as the point of contact of the tangent, other than UC, which can be drawn from U to the conic; and when P is given, Q is found by the line AP. Conversely when Q is given, P is found, and, thence, U follows. As, however, this argument, as we have stated it, involves a tacit reference to the symbols, we may give an alternative one. In fact, the points Q, C are harmonic conjugates of one another in regard to B and U; that is, by perspective from P, if the point of intersection of TU and AC be L, the line BP meets AC in a point which is the harmonic conjugate of L in regard to A and C; in other words BP is the polar line of L; for as L lies on the tangent at P, and on the polar of B, the polar of L passes through P, and through B. We have then to shew that, if Q, U be variable points of the line joining two fixed points B and C, such that B, U are harmonic conjugates of one another in regard to Q and C, then the ranges (Q), (U) are related. For this, take an arbitrary

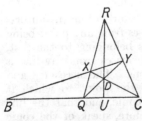

fixed point, R, and an arbitrary line, BXY, passing through B and meeting RC, RQ, respectively, in Y and X; thus Y is a fixed point; if then QY, CX meet in D, the line RD passes through U, and the line BD is a fixed line, meeting YC in the harmonic conjugate of R in regard to Y and C. The range (Q), by perspective from Y, is then related to the range (D) on this fixed line BD, and the range (D), by perspective from R, is related to the range (U). This makes the result clear.

Having proved that the range of points, U, on the tangent CB of the conic, is related to the pencil of lines joining A to the points, P, of the conic; and, then, similarly, that this pencil is related to the range of points, T, on the tangent AB, it follows that these ranges (U), (T) are related. Therein, to the point B, regarded as a point of AB, corresponds the point of contact, C, of the conic with BC, and to B, regarded as a point of BC, corresponds the point of contact, A, of the conic with BA. This is exactly the dual of what was the case when the points of the conic were determined by related pencils.

Examples of the application of the foregoing theorems. We may now regard ourselves as having given the indispensable fundamental propositions of a theory of conics. We proceed to make applications of these, in various directions.

Ex. 1. *The self-polar triad of points for conics through four points; the self-polar triad of lines for conics touching four lines.* It has appeared that if P, P', Q, Q', be four points of a conic, and the three pairs of lines which contain these points intersect respectively in the points O, H and K, then the line HK is the polar

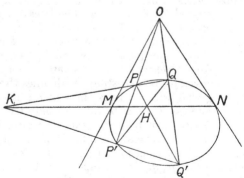

of O; similarly the line OH is the polar of K; and, therefore, the line OK, containing the poles of HK and HO, is the polar of H. For this reason the triad O, H, K is called a self-polar triad, in regard to the conic. The same triad clearly arises for all conics which contain the four points P, P', Q, Q'. We saw that the conics through these four points meet an arbitrary line in pairs of points in involution; this involution has two double points. The conic through the four points and one of these, will not intersect the line again; this line will, therefore, be the tangent of the conic at this point. In other words, two conics can be drawn through four points to have a given line as tangent.

It was remarked that a tangent of the conic meets the curve in two coincident points. It will be found independently that, if this is understood, theorems in regard to four points of the curve continue to hold when these are not all distinct. For instance, if C, A, B be three points of the curve, and a line meet CA, CB respectively in L and M; meet the tangent at C, and the line AB, respectively in T and R; and meet the curve in X and Y, then it can be proved that the pairs L, M; T, R; X, Y belong to an involution. Or again, if A and C be two points of the curve, and a line meet AC in O; meet the tangents at A and C respectively in T and R; and meet the curve in X and Y; then O is one of the double points of the involution determined by the pairs T, R; X, Y. The other double point, U, is the harmonic conjugate of O in regard to T and R, or X and Y. In this latter case, only one conic can be drawn to pass through A and C and have at these points, for tangents, the given lines AT, CR, which shall touch the given line OU. It touches this at U. A degenerate conic having two coincident points at A, and at C, which also has two coincident points on OU, consists of the line OCA taken twice over.

Reciprocally, the aggregate of the tangents of a conic consists of lines, p, whose intersections with two given lines, a, c, form two

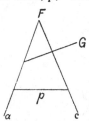

related ranges thereon. The lines a, c are tangents of this conic. If the ranges have the point of intersection, F, of the lines a, c, for a self-corresponding point, they are in perspective, say, from a centre G. In this case we may regard the envelope as consisting of all lines through G, together with all lines through F; it is sometimes said to degenerate into a point pair, F and G. All this has been remarked above (p. 11). In the general case, such a conic envelope can be found having five arbitrary lines, of general position, for tangents. Let four tangents of such an envelope be p, p', q, q'; corresponding to the three ways of dividing these into two pairs, we have three pairs of points of intersection of the four lines. Let o, h, k be the joins of these pairs. Then, as the result dually corresponding to that above, any two of the lines o, h, k intersect in the pole of the remaining line, and form what we may call a self-polar triad of lines in regard to the envelope. There is an infinite number of conic envelopes touching the four lines p, p', q, q'; among these there are three point pairs, the pairs referred to above. The pairs of tangents that can be drawn, from an arbitrary point of the plane, one pair to each conic touching the four lines p, p', q, q', form a pencil in involution; of these pairs, three are those which join the arbitrary point to one of the point pairs referred to, namely, the pairs of points (p, p') and (q, q'); (p, q) and (p', q'); (p, q') and (p', q). This pencil in involution has two double rays; there are thus two conics which can be drawn, to touch the four lines p, p', q, q' and pass through an arbitrary point of the plane; the tangents of these conics at this point are the two double rays of the pencil in involution referred to.

In particular only one proper conic can be drawn to pass through an arbitrary point, O, and touch each of two given lines, which intersect in B, at given points, A, C. The tangent of this conic at O meets the line AC in the harmonic conjugate, with respect to A and C, of the point where the line BO meets AC. A degenerate conic with the definition given, consists of the point B taken twice over, that is of the aggregate of all lines through this point, taken twice over.

Ex. 2. Dual of Pascal's theorem. To Pascal's theorem there is a dually corresponding one, known as Brianchon's theorem. This may be stated by saying that, if a, b, c, a', b', c' be any six tangents of a conic, the three lines joining, the point (b, c') to the point (b', c), the point (c, a') to the point (c', a), and the point (a, b') to the point (a', b), are three lines which meet in a point. Taking the six lines in the order a, b', c, a', b, c', we may speak of the lines which are stated to meet in a point, as those joining opposite angular points of a hexagon, or as diagonals of this.

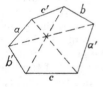

We saw that a tangent of a conic, regarded as a locus of points, may be regarded as meeting the conic in two coincident points. Dually, for a conic envelope, the point of contact of any tangent may be regarded as the intersection of

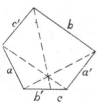

two coincident tangents. Suppose then that, in the general figure for Brianchon's theorem, the two tangents b' and c coincide. We then infer that if a', b, c', a and b' or c, be five tangents of a conic, the point of contact of the last line lies on the line joining the point of intersection (b, c') to the point where the join of the points (a, b'), (a', b) meets the join of the points (a, c'), (a', c). We have given above, in connexion with Pascal's theorem, a construction for the tangent at any one of the five points by which a conic is determined; it may be seen that the present construction, for the point of contact, follows from this by the principle of duality. A particular case, when *b* and *c'* coincide, suggests the result that, if *AB, BC, CD, DA* be four tangents of a conic, the join of the points of contact with *AB* and *CD* passes through the intersection of *AC* and *BD*.

We remarked the existence, given six points of a conic, of sixty Pascal lines. There are similarly, given six tangents, sixty Brianchon points. The properties of the Pascal lines, which are extremely curious, have been much studied. It will be seen below, in Note II, pp. 219 ff., that these properties can be summarised and grasped most easily by regarding the figure as obtained by projection from a figure in space of three, or, better still, of four, dimensions.

Ex. 3. The director conic of a given conic in respect of two arbitrary points. Let *S* be a given conic, and *P*, *Q* be two given points, not lying on the conic, in general. Let any line, *l*, be drawn through *P*, and, then, the line, *m*, be drawn through *Q* which contains the pole of the line *l* in regard to *S*. Thus the line *l* also contains the pole of *m*; and the lines *l*, *m* are a pair of lines, respectively through *P* and *Q*, each of which contains the pole of the other; two such lines are generally spoken of as *conjugate* to one another, in regard to the conic. We prove that the lines *l*, *m* intersect in a point which describes another conic; this conic contains *P*, and contains the points of contact of the two tangents which can be drawn from *P* to touch the given conic *S*; it also contains *Q*, and the two points of contact of the tangents from *Q* to the given conic. Further, the tangents at *P* and *Q*, of this conic, meet in the point, *H*, which is the pole of the line *PQ* in regard to the given conic. For reasons which will appear, this new conic may be called the *director* conic of the given conic, in regard to *P* and *Q*.

In fact, the poles, in regard to the conic *S*, of all lines drawn through *P*, lie upon a line, the polar line of *P* in regard to *S*; and they constitute a range upon this line which is related to the pencil of lines, spoken of, drawn through *P*. This fact is, at once, to be proved directly, by remarking that the pole of any line, *l*, through *P* lies on the polar of *P*, in regard to *S*, and that the harmonic conjugate of this pole, in regard to the two points where the polar of *P* cuts the conic *S*, lies on the line *l*; or the fact follows from what was remarked above, that, in two dually corresponding figures, a range of points, of one figure, is related to the corresponding pencil of lines, of the other figure. Therefore the lines *m*, through the point *Q*, each drawn to the pole of a line *l*, form a pencil related to the pencil of lines *l*; and thus the point of intersection of corresponding lines of these pencils describes a conic containing the points *P* and *Q*. Either of the tangents drawn to the conic *S* from the point *P*, is a particular line *l*, whose pole is its point of contact with *S*; thus the corresponding line *m* contains this point of contact, which is therefore on the conic described. Similarly for the other three points of contact of which we have spoken. If *H* be the pole of the line *PQ*, in regard to *S*, the pole of the line *PH* is on the line *PQ*; this line is then the line through *Q* which corresponds to the line *PH* through *P*; thus, by what was seen above, *PH* is the tangent of the new conic at the point *P*. So *QH* is the tangent at *Q*.

It is easy to see that if the line *PQ* touches the conic *S*, the director conic breaks up into *PQ* and another line.

Ex. 4. *The polars of three points determine a triad of points in perspective with the original triad.* As an application of the ideas involved in the definition of a conic, we may prove that, if *A*, *B*, *C* be three arbitrary points, whose polars, in regard to a conic are the lines *a*, *b*, *c*, respectively, and the points of intersection (*b*, *c*), (*c*, *a*), (*a*, *b*) be, respectively, called *A′*, *B′*, *C′*, then the lines *AA′*, *BB′*, *CC′* meet in a point. The points *A′*, *B′*, *C′* are, respectively, the poles of the lines *BC*, *CA*, *AB*, and the figure may start from these.

Take the points *B*, *C*, *A′*, *C′*; these fix the line *BA*, the polar of *C′*, and fix the line *A′B′*, the polar of *C*. On the line *BA*, take various positions of *A*, to which correspond various positions of *C′B′*, the polar of *A*, and, hence, various positions of *B′*, the pole of *CA*. Let *BB′* and *AA′* meet in *P*. The pencil formed by the various lines joining *A′* to *P*, or, say, the pencil *A′*(*P*), is then related to the range (*A*); this range, determined by the pencil *C*(*A*), of lines which are polars of the various positions of *B′*, is related to the range (*B′*), on the fixed line *A′B′*. For in two dually corresponding figures, a range of points is related to the corresponding pencil of lines, as has been remarked. And 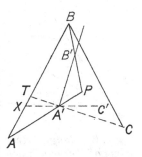 the range (*B′*) is determined by the pencil *B*(*P*). The pencils *A′*(*P*) and *B*(*P*) are thus related, and the point *P* describes a conic. But, when *A* is at *B*, the point *B′* is at *A′*, so that the ray *A′B* of the former pencil corresponds to the ray *BA′* of the latter. The conic which is the locus of *P*, thus breaks up into the line *A′B* and another line. If, however, the lines *C′A′* and *CA′* meet *AB* in *X* and *T*, respectively, it is at once seen that, when *A* is at *X*, the point *P* is at *C′*, and, when *A* is at *T*, the point *P* is at *C*. Thus *P*, *C′*, *C* are in line. And it is this which it was desired to prove. Another proof of this result is given below, Ex. 10.

Ex. 5. *A construction for two points conjugate in regard to a conic.* Two points which are such that the polar of either contains the other are said to be conjugate in regard to the conic. The following simple property seems to deserve remark : If P, X, Y be three points of a conic, the lines PX, PY are met by any line

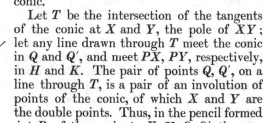

drawn through the pole of the line XY in points which are conjugate in regard to the conic.

Let T be the intersection of the tangents of the conic at X and Y, the pole of XY; let any line drawn through T meet the conic in Q and Q', and meet PX, PY, respectively, in H and K. The pair of points Q, Q', on a line through T, is a pair of an involution of points of the conic, of which X and Y are the double points. Thus, in the pencil formed by lines joining the point P, of the conic, to X, Y, Q, Q', the rays PX, PY are harmonic conjugates in regard to the rays PH, PK. Therefore, on the line TH, the points H, K are harmonic conjugates in regard to Q and Q'. Wherefore the pole of H contains K, or H and K are conjugate points in regard to the conic.

The dually corresponding theorem is that, if x, y, p be three

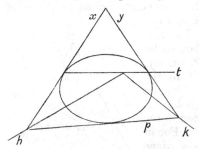

tangents of a conic, and any point of the line, t, which joins the points of contact of x and y, be joined, by lines h and k, to the points where p meets x and y, respectively, then the lines h, k are conjugate to one another in regard to the conic.

The result was remarked by von Staudt (*Geom. der Lage*, 1847, p. 141, No. 253).

Ex. 6. *Range determined by a conic, and by the joins of three points of the conic, upon any line drawn through a fixed point of the*

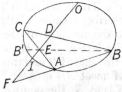

conic. Let A, B, C, O be four fixed points of a conic. Let any line drawn through O meet BC, CA, AB, respectively, in D, E and F, and meet the conic again in I. Then the range D, E, F, I is related to the pencil which the points A, B, C, O subtend at any point of the conic.

For let BE meet the conic again in B'. The range D, E, F, I, by perspective from B, gives the range C, B', A, I on the conic. But, as the lines OI, AC, BB' meet in E, the pairs of points OI, AC, BB' belong to an involution of points on the conic; therefore the range C, B', A, I, of points of the conic, is related to the range of points A, B, C, O, respectively paired with the former in this involution. Wherefore, the range D, E, F, I is related to the pencil of lines joining any point of the conic to the points A, B, C, O; as stated.

Ex. 7. The six joins, of two triads of points of a conic, touch another conic. Let O, I, J and A, B, C be any two triads of points of a conic; then the six joins consisting of BC, CA, AB and OI, OJ, IJ all touch another conic. For, as in the preceding example, let OI meet BC, CA, AB, respectively, in D, E, F; and, let OJ meet BC, CA, AB, respectively, in P, Q, R. The two ranges, D, E, F, I and P, Q, R, J, being, by the preceding example, both related to the points A, B, C, O of the conic, are related to one another. The two lines OI, OJ are thus met in related ranges by the four lines BC, CA, AB, IJ. Therefore, by the definition of a conic, in dual form, these six lines all touch a conic.

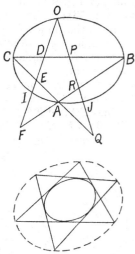

Dually we have the theorem that, if two triads of lines are tangents of a conic, and we take the three intersections in pairs of the lines of a triad, the six points so obtained lie on another conic.

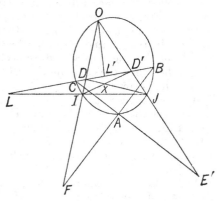

Ex. 8. The pedal line of three points of a conic in regard to three other points of the conic, taken in a particular order. The following result, though somewhat involved, may be proved here, because it depends on the same figure as that just employed. It is of interest later. Let A, B, C and O, I, J be two triads of points of a conic; let IJ meet BC, CA, AB, respectively, in L, M and N. From O are drawn three lines which are the harmonic conjugates, respectively, of OL, OM and ON, in each case in regard to OI and OJ. If these lines be, respectively, OL', OM', ON', then the pairs OL, OL'; OM, OM', and ON, ON' are in involution, of which the double

rays are OI, OJ. Supposing L', M', N' taken, respectively, upon BC, CA, AB, the theorem to be proved is, that L', M', N' are in line.

Let OI meet BC, CA, AB, respectively, in D, E, F, and OJ meet these, respectively, in D', E', F'. By what was seen in Ex. 6, the ranges D, E, F, I and D', E', F', J are related. Wherefore, the three points of intersection (ID', JD), (IE', JE), (IF', JF) are in line (Vol. I, p. 53); let these points be named, respectively, X, Y and Z. Of these, the point X is evidently on the line OL', being the harmonic conjugate of O, in regard to the point L' and the point where OL' meets IJ. Similarly, Y and Z are, respectively, on OM' and ON', and there is a corresponding harmonic relation. If then the line XYZ meet IJ in T, the points L', M', N' lie on a line through T, this being the harmonic conjugate of the line TIJ in regard to the lines TO and $TXYZ$.

Ex. 9. The Hessian line of three points of a conic. The Hessian point of three tangents of a conic. If the tangents at the points

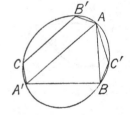

A, B, C, of a conic, meet BC, CA, AB respectively in P, Q, R, then these points are in line.

By Pascal's theorem, if A, B', C, A', B, C' be six points of a conic, the three points of intersection $(AB', A'B)$, $(BC', B'C)$, $(CA', C'A)$ are in line. Suppose now that B' coincides with A, and A' with C, and C' with B. Then, instead of the lines AB', BC', CA', there will be the tangents of the conic at A, B and C, respectively; and, instead of the lines $A'B, B'C, C'A$, there will be the lines BC, CA, AB, respectively. The theorem in question thus expresses the fact that Pascal's theorem continues to hold when these coincidences occur. We can, in fact, prove it directly in the same way as Pascal's theorem. For, going back to the original definition of a tangent, we see that the theorem expresses that if two ranges of points of the conic be related to one another by the condition that the points A, B, C, of the one, correspond, respectively, to the points C, A, B, of the other, then the common corresponding points of these ranges are on a line containing the points, P, Q, R, where the tangents at A, B, C meet, respectively, the opposite joins (cf. Vol. I, p. 174, Ex. 5). We shall speak of the line PQR as the Hessian line of the triad A, B, C in regard to the conic.

If the tangents of the conic at B, C meet in A', the line AA' is the polar of P. The three lines such as AA' thus meet in a point, which is the pole of the line PQR.

Dually, we have the result that, if a conic touch the joins VW, WU, UV, of any three points, U, V, W, respectively, in A, B, C, then the lines AU, BV, CW meet in a point. We shall call this the Hessian point of the three tangents in regard to the conic.

If BC meet VW in D, then D is the pole of UA. Thus the three points such as D are in line.

Ex. 10. *Polar triads in regard to a conic. Hesse's theorem.* The preceding result is a particular case of another, of which a proof has already been given (Ex. 4, above), which we may state as follows: The polar lines, in regard to a conic, of any three points A, B, C, meet the joins BC, CA, AB, respectively, in three points which are in line. If the polars of B and C meet in A', the polars of C and A meet in B', and the polars of A and B meet in C', the result will follow, by Desargues' theorem of perspective triads, if we prove that the lines AA', BB', CC' meet in a point. But if BB' and CC' meet in D, and we notice that B' is the pole of CA and C' the pole of AB, this is the same as saying that, if A, B, C, D be any four points such that one pair of joins of these, AC and BD, be conjugate lines in regard to a conic, and also another pair of joins of these, AB and CD, be conjugate lines in regard to this conic, then also the third pair of joins of these, BC and AD, are conjugate lines in regard to this conic. This result may be called Hesse's theorem; it is of importance in the sequel.

In dual form, this theorem is that, if any four lines a, b, c, d be divided into pairs in each of the three possible ways, and the points,

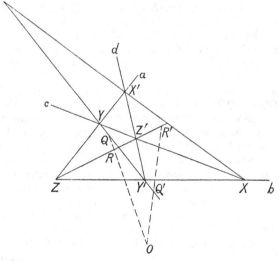

Y, Y', obtained by the intersection of the pair a, c, and by the intersection of the pair b, d, be conjugate points in regard to

conic; and the same be true of the points, Z, Z', obtained by the intersection of the pair a, b, and by the intersection of the pair c, d: then the points, X, X', the intersections, respectively, of b and c, and of a and d, are also conjugate points in regard to the conic. We prove this by shewing that the points in which the conic meets the line XX' are harmonic conjugates of one another in regard to X and X'.

Let the conic meet the line YY' in Q and Q'; since the polar of Y passes through Y', the points Q, Q' are harmonic conjugates of one another in regard to Y and Y'. The points, R and R', in which the conic meets the line ZZ' are, similarly, harmonic conjugates of one another in regard to Z and Z'. Let QR and $Q'R'$ meet in O. As the conics through the four points Q, R, Q', R' meet the line XX' in pairs of points belonging to an involution, the given conic will be shewn to meet XX' in points harmonic in regard to X and X', if there be found two other conics, through the four points Q, R, Q', R', having this property; the points X, X' will then be the double points of the involution. By the original definition of the harmonic relation, the line pair QQ', RR', or YY' and ZZ', is one such conic. In fact the line pair OQR, $OQ'R'$ is another such conic. For the three line pairs joining O to the point pairs Y, Y'; Z, Z' and X, X' are in involution (Ex. 1, above), and the lines OQR, $OQ'R'$ are harmonic in regard to OY, OY', and also in regard to OZ, OZ', as we have seen; they are, therefore, the double rays of this pencil in involution of centre O, and are thus, harmonic in regard to OX and OX'. Thus OQR, $OQ'R'$· are, as stated, a second conic through the four points Q, R, Q', R' which cuts XX' in points harmonic in regard to X and X'. The result required is thus obtained.

We have seen, in Ex. 1, above, that the tangent pairs, from the point O, to conics which touch the four lines a, b, c, d, are a pencil in involution, of which OY, OY' are a pair of rays, as are OZ, OZ'; the lines OQR, $OQ'R'$ being, as we have just remarked, the double rays of this pencil, are then harmonic in regard to the tangents from O to any such conic. Now, two points, which are harmonic in regard to the two points in which their join meets a conic, are conjugate points in regard to the conic; and, dually, two lines, which are harmonic in regard to the tangents drawn to a conic from their intersection, are conjugate lines in regard to the conic. Thus, the lines OQR, $OQ'R'$ are conjugate lines in regard to any conic touching the four lines a, b, c, d. The lines YY' and ZZ' have also been seen (Ex. 1, above) to be conjugate to one another in regard to any such conic. The line pairs OQR, $OQ'R'$ and YY', ZZ' (or QQ', RR') are two of the three line pairs containing the points Q, R, Q', R'; by the dual of the theorem proved in this example, it follows, since two of these line pairs are conjugate lines in regard to any particular conic, Σ, touching a, b, c, d, that the third line pair, QR', $Q'R$, containing these points, are also conjugate in regard to Σ. We have thus proved a theorem which may be stated thus: Let a, b, c, d be any four tangents of a conic, Σ; let two (and, therefore, the third) of the point pairs of intersection of these tangents (Y, Y' and Z, Z') be conjugate points in regard to another conic, S. There is then a set of four points of the conic S (namely Q, R, Q', R') such that each line pair, containing these, is a pair of conjugate lines in regard to Σ.

Ex. 11. *Outpolar conics.* Let Q, R, Q', R' be four points of a conic, S, which are such that two (and, therefore, the third) of the line pairs containing these, say QR, $Q'R'$ and QQ', RR', are pairs of conjugate lines in regard to another conic, Σ. Such a figure is not possible for any two conics S, Σ, arbitrarily given. When it is possible, we shall speak of the conic S as being *outpolar* to Σ. Dually, when there are four tangents of Σ whose three point pairs of intersection are pairs of conjugate points in regard to S, we shall speak of Σ as being *inpolar* to S. The result obtained at the conclusion of the preceding example may then be stated by saying that when Σ is inpolar to S, then S is outpolar to Σ. By dual reasoning it follows that, when S is outpolar to Σ, then Σ is inpolar to S.

Proceeding now with the hypothesis made at the beginning of this example, that the line pairs QR, $Q'R'$ and QQ', RR', are conjugate lines in regard to Σ, take the polar line of Q, in regard to Σ, and let this meet the conic, S, containing the points Q, R, Q', R', in the points M and N. On MN take the pole, N', of QM, also in regard to Σ, so that M is the pole of QN'. The lines QM, $R'N'$ are then conjugate in regard to the conic Σ; and the lines QN', $R'M$ are also conjugate in regard to Σ. Now, when any lines are drawn through Q, their poles, in regard to Σ, lie on the polar line MN of Q in regard to Σ; and

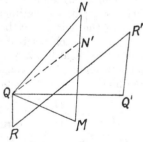

the lines joining R' to these poles form a pencil related to the range of these poles, and therefore related to the pencil of lines drawn through Q. Thus the pencil $Q(M, N', Q', R)$ is related to the pencil $R'(N', M, R, Q')$, and, hence, to the pencil $R'(M, N', Q', R)$, for the lines QQ', $R'R$ were given to be conjugate lines in regard to Σ, as were QR and $R'Q'$. It follows that the six points Q, Q', R, R', M, N' lie on a conic; this is, therefore, the same as the conic, S; and the point N', not coinciding with M in general, coincides with N. The points Q, M, N' were constructed to form a self-polar triad in regard to Σ; the points Q, M, N thus form such a triad. We have, therefore, shewn that when the conic S is out-polar to Σ, as containing four points whose line pairs of joins are conjugate pairs in regard to Σ, then a triad of points can be found on S which are a self-polar triad in regard to Σ. And one point of this triad may be taken to be any one of the four given points of S.

Conversely, let a triad of points of S, say Q, M, N, form a self-polar triad in regard to Σ; take, upon S, two other quite arbitrary points, R and R'; draw, also, the line QQ' through Q which is conjugate to the line RR' in regard to Σ, and draw the line $R'Q'$ through R' which is conjugate to QR in regard to Σ; let these lines meet in Q'. It can then be proved, as above, that the point Q' lies on the conic S. There are then four points, Q, R, Q', R', of the conic S, such that the lines $QR, Q'R'$ are conjugate in regard to Σ, as also are the lines QQ' and RR' (and, therefore, also QR' and $Q'R$). The conic S is thus outpolar to Σ. Therefore, by what was proved above, if the polar of R in regard to Σ be taken, this will meet S in two points which, with R, form a triad of points of S which is self-polar in regard to Σ. The point R was an arbitrary point of S; such triads exist, therefore, in infinite number, of which one point is arbitrary.

Hence, if two conics S, Σ be so related that there is one triad of points of S which are a self-polar triad in regard to Σ, there is an infinite number of such triads, one of the three points being arbitrary. There is then also an infinite number of sets of four points of S, which are such that each of the line pairs containing them is a pair of conjugate lines in regard to Σ; of such a set, three of the points may be taken arbitrarily upon S.

If two of such a set of four points upon S be conjugate to one another in regard to Σ, then the set consists of a triad which is self-polar in regard to Σ (consisting of these two points and another), together with a fourth point. For let A, B, C, D be such a set, the lines AB, CD being conjugate in regard to Σ, as also AC, BD, while B and C are conjugate in regard to Σ; and suppose A, B, C are not a self-polar triad in regard to Σ. Then the pole of AB is some point of CD, say C_1, and C_1 is not at C. The polar of C_1 being AB, and the polar of C passing through B, it follows that B is the pole of the line CC_1, or CD. In a similar way C is the pole of BD. Thus B, C, D are a self-polar triad in regard to Σ. Conversely, if B, C, D be points of S forming a self-polar triad in regard to Σ, and A be any other point of S, the four points A, B, C, D have the property that every line pair containing them consists of conjugate lines in regard to Σ.

By what has been said, it follows, also, that if there are three points of S forming a self-polar triad in regard to Σ, then, the conic Σ being inpolar to S, there are sets of three tangents of Σ forming a self-polar triad of lines in regard to S.

An important consequence of what has been said is, that if S and S' be two conics which are both outpolar to a conic Σ, then the four common points of S and S' are such a set of four points as has been spoken of: every line pair which contains these four points consists of lines which are conjugate to one another in regard to Σ. And, thus, every conic through the four common points of S and S' is also outpolar to Σ.

For, let A, B, C be three points common to S and S'; if these form a self-polar triad in regard to Σ, then these, with the fourth intersection of S and S', form such a set of four points; and, every conic through A, B, C is outpolar to Σ. When A, B, C do not form such a self-polar triad, by drawing, from B, the line BD, conjugate to AC, in regard to Σ, and drawing, from C, the line CD, conjugate to AB, in regard to Σ, we determine a point D. By what has been proved above, this point D lies both on S and S'. There is a dually corresponding theorem in regard to the four common tangents of two conics Σ, Σ' which are both inpolar to a conic S.

Ex. 12. *Two self-polar triads of one conic are six points of another conic.* One incidental consequence of the theory of the preceding example is, that if two triads, A, B, C and A', B', C', be both self-polar in regard to a conic, Σ, then the six points A, B, C, A', B', C' lie on another conic. For the conic, S, which is drawn to contain A, B, C and A', B', is outpolar to Σ, as containing A, B and C. Therefore, by what has been shewn, the polar line of A', in regard to Σ, meets S in two points which, with A', are a self-polar triad in regard to Σ. One of these points is B'; the other is on $B'C'$, and is the pole of $A'B'$ in regard to Σ; it is therefore C'.

The direct proof, which is equivalent to a repetition, may be given. The lines AB, AC, AB', AC' are evidently conjugate, in regard to Σ, respectively, to $A'C$, $A'B$, $A'C'$, $A'B'$. Thus the pencil $A(B, C, B', C')$ is related to the pencil $A'(C, B, C', B')$, and hence, also, to the pencil $A'(B, C, B', C')$. Thus the six points lie on a conic.

By the dually corresponding proof, we can shew that the six lines BC, CA, AB, $B'C'$, $C'A'$, $A'B'$ all touch the same conic.

We shall (Ex. 21) prove, conversely, that, if A, B, C, A', B', C'

3—2

be any six points of a conic, there exists another conic in regard
to which both A, B, C and A', B', C' are self-polar triads. By combining this with the theorem here, we shall thus have another
proof of a theorem given above (Ex. 7), that if A, B, C and A',
B', C' be points of a conic, there is another conic which touches all
of BC, CA, AB and $B'C'$, $C'A'$, $A'B'$.

Ex. 13. *Particular cases of outpolar, or of inpolar, conics.* A
particular case of a conic, S, outpolar to a proper conic, Σ, is a
pair of lines which are conjugate in regard to Σ. A particular case
of a conic, Σ, inpolar to a proper conic, S, is a pair of points which
are conjugate in regard to S. In the general case, while the conic S
is regarded as the locus of its points, the conic Σ is regarded as the
envelope of its tangents; thus, a better statement of the latter case
is, that the degenerate conic, Σ, consists of all lines through one of
the two points conjugate in regard to S. A line pair may also be
spoken of as inpolar to a conic, S, when the lines intersect at a
point lying on S. But this point is to be regarded as a point pair
of which the points coincide; the conic Σ consists then of all lines
through this point. Similarly a line, taken twice over, may be
spoken of as outpolar to a conic Σ when the line touches Σ.

These statements may be verified geometrically. Further con-
firmation will arise below when the matter is treated with the help
of the algebraic symbols (Chap. III, Ex. 24).

Ex. 14. *The joins of four points of a conic and the intersections
of the tangents at these points. The common points, and the common
tangents, of two conics.* We have seen (Ex. 1) that the intersections
of the three line pairs, which contain four points of a conic, form a
self-polar triad in regard to the conic. Dually, the lines which join
the three point pairs, through which four tangents of the conic
pass, are a self-polar triad of lines in regard to the conic. When
the four points are the points of contact of the four tangents considered, the triad of lines consists of the joins of the triad of points.
In other words, the tangents at four points of a conic meet in pairs
on the joins of the three points which form the common self-polar
triad of all conics passing through the four points. We have seen,
in Ex. 2, that, effectively, the result follows directly from Brianchon's theorem.

Let the line pairs containing the points A, B, C, D of a conic
intersect, respectively, in E, F and G. The tangents of the conic at
A and B meet in the pole of the line AB; as AB passes through
one of the points E, F, G, say, through F, the pole of AB lies on
the polar of F, that is, on the line EG. In the same way, the point
of intersection of the tangents at any two of the points A, B, C, D
is on one of the lines FG, GE, EF.

We have seen that, if we have two conics, the intersections of the line pairs passing through their intersections, form a self-polar triad of points for both conics. Dually, the joins of the point pairs, through which the common tangents of the two conics pass, form a self-polar triad of lines for the two conics (Ex. 1, above). These lines are in fact the joins of the points spoken of. In other words, through a point of intersection of two complementary chords common to two conics, there pass two lines each containing the intersection of two common tangents of the conics, and also the intersection of the other two common tangents; this line passes through the intersection of another pair of complementary chords common to the two conics.

To see that this is so, let us speak of two points P, P' as being harmonic images of one another in regard to a given point F, and a given line EG, when they are harmonic conjugates of one another, in regard to F, and the point where FP meets EG. Then either of P, P' determines the other; and if Q' be the harmonic image of another point Q, the lines $PQ, P'Q'$ intersect on EG. We may then regard a conic, for which E, G, F is a self-polar triad, as consisting of pairs of points, P, P', which are harmonic images of one another in regard to F and EG, the tangents of the conic at P and P' intersecting on EG. And, if E, G, F be the intersections of the line pairs through the common points of two conics, the harmonic image of a common tangent of the two conics is another common tangent of these conics, meeting the former on the line EG.

Ex. 15. *The polar lines of any point in regard to the conics through four given points all pass through another point.* For if the polars of a point, P, taken in regard to two of these conics, meet in a point P', the points P, P' are harmonic conjugates of one another in regard to the two points in which the line PP' meets either of these conics; they are, therefore, the double points of the involution determined on this line by the two pairs of intersections of this line with these two conics. As all the conics meet this line in pairs of points belonging to this involution, the points P, P', being harmonic conjugates in regard to any pair of the involution, are conjugate points in regard to any one of the conics. Thus the polar of P in regard to any one of the conics passes through P'.

In particular, if E be the intersection of one of the line pairs through the four points, the lines EP', EP are harmonic conjugates of one another in regard to this pair of lines. Thus, if F be the intersection of another of the line pairs through the four points, the point P' may be constructed from P by joining P to E and drawing a fourth harmonic line through E, and joining P to F and drawing another such fourth harmonic line through F. The point P' is the intersection of these. It lies on the similar fourth harmonic line which can be drawn through the intersection of the remaining line pair containing the four points.

Dually, the poles of any line in regard to all conics touching four given lines, lie on another line. This can be constructed by joining points determined as fourth harmonic points, on the lines containing the point pairs through which the four given lines pass.

Ex. 16. *A point to point correspondence in regard to four given points.* Of the two points P, P' of the preceding example, either determines the other without ambiguity, when the four points A, B, C, D are given; and the relation is reciprocal or involutory, in the sense that, if P take the position of P', then P' takes the position of P. To this statement, however, there is an exception: Let E, F, G be the intersections of the line pairs containing A, B, C, D. Then when P is on the line FG, the point P' is at E; thus, to the position E, for P', the corresponding position of P is undetermined, being anywhere on the line FG. Similarly when P' is at F or G.

It can be shewn that, if P describes any line, then P' describes a conic passing through E, F, G; and, conversely, if P' describes any conic passing through E, F, G, then P describes a line. Evidently, conics passing through three fixed points E, F, G, are in some sense analogous to lines; in that any two such conics have one point of intersection, beside E, F, G, while such a conic can be drawn through two arbitrary points, other than E, F, G. The relation now in question enables us to pass from theorems relating to lines to theorems relating to such conics, or conversely.

To prove the statement made: when P describes any line, the pencils of lines joining E and F to P, or say $E(P)$ and $F(P)$, are related. As, however, the lines EP, EP' are harmonic in regard to the line pair intersecting in E, the pencils $E(P')$, $E(P)$ are related. Similarly, so are the pencils $F(P')$, $F(P)$. Thus the pencils $E(P')$, $F(P')$ are related. Thus, P' describes a conic containing the points E and F. This conic contains G; for this is the position of P' when P is on the line EF, at the point where this is met by the line on which P lies.

Now suppose that any points E, F, G are given, and also any other two points I, J. We shew that points A, B, C, D can be found so that, (a), the points E, F, G are the intersections of the line pairs containing A, B, C, D, (b), the points I, J are points corresponding to one another, in regard to A, B, C, D, in the sense just considered. The correspondence in question can then be regarded as determined by an arbitrary triad of points, E, F, G, and one given pair of corresponding points.

In fact two lines, FBA, FCD, are determined by the conditions that, they shall be harmonic conjugates of one another in regard to the given lines FE, FG, and shall also be harmonic conjugates of one another in regard to the given lines FI, FJ. And two lines, GDA, GCB, are determined by the conditions that, they shall be harmonic in regard to GE, GF, and also harmonic in regard to GI, GJ. These lines determine four points A, B,

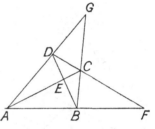

C, D. We have to shew that AC, BD meet in E, and are harmonic in regard to EI, EJ. In fact, AC and BD must meet on the line through G, which is the harmonic conjugate of GF in regard to the lines GCB and GDA; and they must, likewise, meet on the line through F which is the harmonic conjugate of FG in regard to the lines FBA and FCD; therefore AC, BD meet in E. And then it follows, by what was said above, that EC, EB are harmonic in regard to EI and EJ.

Ex. 17. *Four conics can be drawn through two given points to touch three given lines.* We have proved above that two conics can be drawn through a given point to touch four given lines. We consider now the conics that can be drawn through two given points to touch three given lines.

For this, notice, first, that if FM, FN be the tangents drawn, from a point F, to a conic, of which I, J are two points, and C be the pole of the line IJ, then the involution determined by the two pairs of rays, FM, FN and FI, FJ, has FC for one double ray. In fact, if MN and IJ meet in O, this is the pole of the line FC, and its harmonic conjugates, both in regard to the pairs I, J, and in

regard to the pair M, N, lie on FC. Thus the lines FO, FC are
harmonic, both in regard to FM, FN, and in regard to FI, FJ.

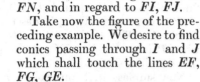

Take now the figure of the preceding example. We desire to find conics passing through I and J which shall touch the lines EF, FG, GE.

One conic can be drawn through I and J, to have CI, CJ as the tangents at these points, which shall touch FG. (Another degenerate conic which contains two coincident points of CI at I, two coincident points of CJ at J, and two coincident points of FG, consists of the line IJ, taken
twice over. See Ex. 1, above.) This conic does, in fact, touch FE and GE, and is therefore such a conic as is desired. To prove that this conic, which touches FG, also touches FE, we use the remark made above, shewing that the involution which is determined by the pair of lines FI, FJ, and the double ray FC, contains the lines FE, FG

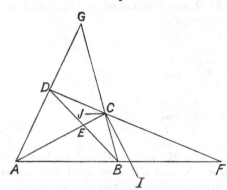

as a pair. For this we have only to notice that FC, FB are harmonic in regard to FE, FG, and also harmonic in regard to FI, FJ (by the property of I, J in regard to A, B, C, D); the lines FC, FB are thus the double rays of an involution of which FI, FJ and FE, FG are pairs. This is then the involution spoken of. By a similar argument the conic described touches GE.

Another conic through I, J touching EF, FG, GE, is that described to touch DI, DJ, respectively, at I, J, and to touch FG. A third touches AI, AJ at I and J; and a fourth touches BI, BJ at I and J.

No other conics can be drawn through I, J to touch EF, FG, GE. For, by the remark made above, if FE, FG be the tangents from F to a conic which contains I and J, one of the double rays of the pencil in involution determined by the pairs, FI, FJ and FE, FG, must contain the pole of IJ in regard to this conic. The pole of IJ must then lie either on the line FAB or the line FCD. It must, similarly, lie either on *GCB* or on *GDA*. It must therefore be at one of *A, B, C, D*.

The dual of this proposition is that four conics can be drawn through three given points to touch both of two given lines. Of this a proof, which then also proves the direct proposition, follows at once from the correspondence explained in the preceding example (Ex. 16). This correspondence transformed conics passing through three given points into lines, and lines into conics passing through these three given points. It therefore changes a conic which touches a line into a line touching a conic, the two common points of these being coincident in both cases. Thus a conic passing through the three given points and touching two given lines, becomes a line touching two conics which pass through the three given points. The result to be proved is, therefore, that, two conics have four common tangents. This we know, as the dual of the result assumed (Preliminary, above, p. 9) that two conics have four common points.

Ex. 18. *The poles of a line, in regard to conics with four common points, lie on a conic.* We have remarked that the poles of a line in regard to conics with four common tangents lie on a line. We may, however, consider the poles of a line in regard to conics with four common points. These lie on another conic, which is of importance. Of this conic eleven points can be specified at once, from the figure; it is therefore sometimes called the eleven-point conic.

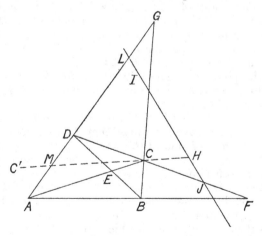

Let *A, B, C, D* be the four common points of the conics considered. These determine, on the given line, an involution of pairs of points; let *I, J* be the double points of this involution. Let *E, F, G* be the intersections of the line pairs containing the four given points. A conic can be drawn through the five points *E, F,*

G, I, J; we shew that this contains the pole of the given line in regard to every conic through the four given points.

For, I, J are harmonic conjugates in regard to the points in which any particular conic, S, through A, B, C, D, meets the line IJ; the polars of I and J, in regard to this particular conic S, thus pass, respectively, through J and I. Therefore, if these polars meet in T, the points T, I, J are a self-polar triad in regard to S. We have shewn that E, F, G are a self-polar triad in regard to any conic S (Ex. 1). We have also shewn that any two triads of points, which are both self-polar in regard to a conic, lie on another conic (Ex. 12). Thus T lies on the conic containing I, J, E, F, G. But T is the pole IJ in regard to the conic S. This proves the statement made.

Now, let the line IJ meet AD in L; and let M be the harmonic conjugate of L in regard to A and D. Further, let MC meet the line IJ in H, and let C' be the harmonic conjugate of C in regard to M and H. If then we consider the particular conic through A, B, C, D which contains the point C', the polar of M in regard to this conic contains both the point L and the point H. Thus M is the pole of the line IJ in regard to this particular conic. Wherefore M is on the conic obtained above, which is the locus of the poles of the line IJ in regard to all conics through A, B, C, D. By considering the intersections of the line IJ with the joins of every pair of the four points A, B, C, D, as we have just considered the intersection L with the join AD, and taking on each join the fourth harmonic in regard to the two points, as we have taken M on AD, we find in all six points of the conic in question. And we found that it contained the five points E, F, G, I, J.

Dually, the polars of a point, in regard to all conics touching four given lines, touch another conic, which will also be found to be of importance (p. 90, below).

Ex. 19. *Two conics of which the tangents at their common points meet, in fours, at two points.* Suppose that the four points common to two conics are I, J and H, K; and that the tangent of one conic, at the point H, passes through the intersection of the tangents of the other conic at the points I and J. For distinctness, let the first conic be called Φ, and the second be called Ω, and let the tangents of Ω at I and J meet in O. It can then be shewn that the tangent of the first conic, Φ, at the point K, also passes through O, which is then a point through which four of the tangents of the conics pass; and that the tangents of the second conic, Ω, at the points H, K, as well as the tangents of the first conic, Φ, at I and J, all pass through another point.

Evidently, when the conic Ω is given, and the points I, J, H, K thereon, the conic Φ may be defined as that passing through the

points I, J, H, K and having HO for its tangent at H. We obtain
it somewhat differently.

Let Ω be any given conic, I, J, H, K any points thereon, the
tangents at I, J meeting in O. Draw through O any line meeting
Ω in P and P'; let $PH, P'K$ meet in Q, and
$PK, P'H$ meet in R. We consider the locus
of the point Q; which will be found to contain
the point R. The pairs of points P, P' of the
conic forming an involution thereon, the
pencils of lines, $H(P), K(P')$, with centres at
H and K, are related. This shews that the
locus of Q is a conic, say Φ, passing through
H and K; and, as P, P' are interchangeable,
this conic Φ contains R. When P' is at H,
the line HP becomes the line HO, and the line
KP' becomes KH; thus, HO is the tangent of
the conic Φ at H. When P is at K, the line

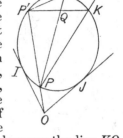

HP becomes the line HK, and the line KP' becomes the line KO,
which is thus the tangent of the conic Φ at K. When P' is at I,
so also is Q. The conic Φ thus passes through I; and, similarly, it
passes through J.

Now let T be the intersection of the tangents of the conic Φ at
I and J. We shew that the tangents
of Ω at H and K pass through T. Let
HO meet IJ in F, and TH meet IJ
in G. The point F is on the polar of
T in regard to Φ, so that its polar
contains T; the polar of F also passes
through H, since the tangent of Φ at
H passes through F. Thus the line
THG is the polar of F in regard to Φ;
and G is the harmonic conjugate of F
in regard to I and J. If, however, the
tangent of Ω at H meet IJ in G', the
polar of G', in regard to Ω, passes
through H, and passes through O,

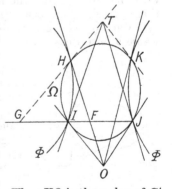

which is the pole of IJ in regard to Ω. Thus HO is the polar of G'
in regard to Ω. Wherefore G and G' are the same, and TH is the
tangent of Ω at H. So TK is the tangent of Ω at K.

It appears, thus, that the relation of the conics Φ, Ω is mutual,
and either may be defined from the other. And the conic Φ may,
equally, be defined by drawing variable lines through the point T
to meet the conic Ω, say, in X and X', and, then, finding the locus
of the intersection of the lines IX and JX'.

In the definition of the conic Φ, from Ω, the points Q, R of Φ are conjugate to one another in regard to all conics through the points P, P', H, K (Ex. 1, above), and the line QR, which is the polar of the intersection of the lines PP' and KH, contains the intersection, T, of the tangents of the conic Ω at H and K. The two conics, Ω, Φ, are, therefore, such that any line drawn through T meets one of the conics, say Φ, in two points which are harmonically conjugate in regard to the two points in which the line meets the other conic Ω. The same is therefore true of any line drawn through the point O.

If Ω be given, and, on any line drawn through the intersection, O, of the tangents of Ω, at any two points, I, J, of this, there be taken two points, A, B, conjugate to one another in regard to Ω, then any conic, Φ, through the four points I, J, A, B, is related to Ω as in the above. In fact, IB, JB pass through the further intersections, C, D, of Ω with AJ, AI, respectively, and CD passes through the intersection of the tangents of Φ at I and J. Or an independent proof may be given.

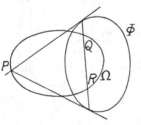

Ex. 20. *A conic derived from two given conics.* In connexion with the last example, the following remarks, though an anticipation of results given below (in Chap. III, p. 125), may be made.

Given any two conics Ω, Φ, we may consider all lines of the plane whose intersections with one of the conics are harmonic conjugates in regard to the intersections of the line with the other conic. Of such lines, two can be drawn through any point, P, of one of the conics, say Ω; those, namely, joining P to the points, Q, R, in which this conic, Ω, is met by the polar of P in regard to Φ. It may, therefore, be expected that the lines in question touch a conic; as will be found to be the case (Chap. III, below). If this be assumed, it is at once clear that this third conic touches the eight lines which are the tangents of the given conics, Ω, Φ, at their four intersections. The two particular conics, Ω, Φ, considered in the preceding example, are so related that this third conic, regarded as an envelope, degenerates into a point pair, consisting of the points O and T. It can be shewn, further, that the points of contact of these particular conics Ω, Φ with their four common tangents, lie on two lines, four on each; and that the tangents, from any point of either of these lines, drawn to Φ, are harmonic conjugates in

regard to those drawn to Ω. In general, a point of a plane from
which the tangents to two general conics form such a harmonic
pencil, describes a conic passing through the eight points of contact
of these given conics with their four common tangents.

*Ex. 21. To construct a conic in regard to which two triads of
points of a given conic shall both be self-polar.* Let A, B, L and
C, D, M be any two triads of points lying on a conic.

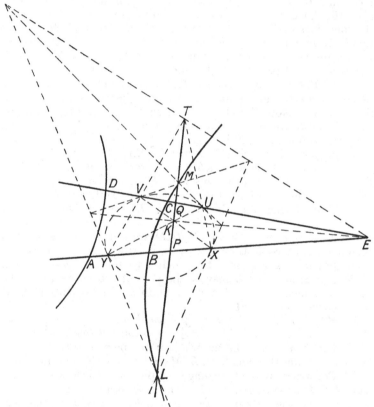

Let the lines AB, CD meet in E, and the line LM meet these
lines, respectively, in P and Q. There exist, on AB, two points,
X, Y, which are harmonic conjugates both in regard to A, B and
in regard to P, E. Any conic for which L, A, B is a self-polar
triad must necessarily meet AB in points which are harmonic conjugates in regard to A and B; for any conic for which L, A, B and
M, C, D are self-polar triads, the intersection, E, of AB and CD,
must be the pole of LM; any such conic must, therefore, meet AB

in points which are harmonic conjugates in regard to P and E. The conic we seek to construct must then, if it exist, pass through X and Y. Two other points through which such a conic must pass are the points, U, V on CD, which are harmonic conjugates both in regard to C, D and in regard to Q, E. It follows from the fundamental properties of the harmonic relation that the lines XU, YV meet on the line LM, say, at T; and, also, that the lines XV, YU meet on LM, say, at K. Considering any four lines, we have seen (Ex. 10, above) that if, of the three point pairs through which all the four lines pass, two pairs consist each of points which are conjugate to one another in regard to a conic, so does the third pair. In the present case, for the four lines TUX, TVY, UKY, VKX, the three point pairs through which all the lines pass are X, Y; U, V and T, K; of these, the two former consist of conjugate points in regard to the given conic containing the two triads L, A, B, M, C, D; wherefore K and T are conjugate in regard to this conic. Thus T and K are harmonic conjugates in regard to L and M. The harmonic conjugate of T in regard to the points U and X is, however, the point in which UX is met by EK. Thus it follows that the lines MU, LX meet on EK; as do, for a similar reason, MV, LY. In the same way, the lines VM, LX meet on ET, as do UM and LY. Whence we see that a conic can be described through X, Y, U, V having MU, MV and LX, LY as tangents (Ex. 14, above).

For this conic, L, A, B and M, C, D are self-polar triads. It is therefore such a conic as we desired to construct. And there can be no other; for we have seen that such a conic must pass through X, Y, U, V; and, that L may be the pole of AB in regard to the conic, its tangents at X, Y must meet in L; its tangents at U, V must likewise meet in M.

Ex. 22. *To construct a conic for which a given triad of points shall be a self-polar triad, and a given point and line shall be pole and polar.* Let L, A, B be the triad, M the point and CD the line. Draw any conic through L, A, B, M; let it cut the line in points C, D. Construct, by the foregoing example, the conic having L, A, B and M, C, D as self-polar triads. This is such a conic as is desired. It is, in fact, the only one; for, though C, D may vary, they are pairs of the involution determined on the line CD by all conics through L, A, B, M; and the points U, V, of the conic described, are unique, being the double points of this involution.

Ex. 23. *To construct a conic for which a given triad of points shall be self-polar, to pass through two given points.* Let A, B, C be the triad, and I, J the two other points. There exists a point, D, such that the lines BC, AD, the lines CA, BD, and the lines AB, CD, meet the line IJ in three pairs of points of an involution of which

I, J are the double points. This is the particular case of Hesse's theorem (Ex. 10, above), which arises when the conic, there referred to, becomes a point pair; it is obvious at once, by finding D, so that BD and CA meet IJ in points harmonic in regard to I, J, and that CD and AB meet IJ in such points; then I, J are the double points of the involution determined on the line IJ by the joins of the points A, B, C, D, so that AD and BC meet IJ in points harmonic in regard to I and J.

There is then, by the preceding example, one conic having A, B, C for a self-polar triad, and having the point D and the line IJ as pole and polar of one another. For this conic the points A and D are the poles, respectively, of BC and IJ, and the line AD is the polar of the point where BC meets IJ. Thus the conic meets the line IJ in points which are harmonic conjugates in regard to the points where IJ is met by BC and AD. There is a similar statement for CA and BD, and for AB and CD. The conic, therefore, passes through I and J, and is the conic we desired to construct. The tangents at I and J are the lines DI and DJ. The reader will compare this work with that of Ex. 17.

Ex. 24. *The polar reciprocal of one conic in regard to another. A particular case.* It follows from what has been said above in regard to duality with respect to a conic, and the dual definition of a conic (p. 22), that if we have two conics, S and Ω, and take the polars, in regard to Ω, of all the points of S, these polars touch another conic, say S'. Any two points of S thus give two tangents of S', and the polars, in regard to Ω, of the points of their join, become the lines passing through the point of intersection of these two corresponding tangents of S'. In particular the pole, in regard to Ω, of any tangent of S, is a point of S'. Thus S can be obtained from S' by the same rule as was S' from S. Two such conics S, S' are called polar reciprocals in regard to Ω. Conversely, it will be seen below, in Chap. III, that, given two arbitrary conics S, S', there are, in general, four conics Ω, in regard to which S and S' are polar reciprocals.

Now consider two conics, S and S', which are such that there is a triad of points of S, say A, B and C, for which the joins BC, CA, AB are tangents of S'. If then we draw from any point, P, of S, the two tangents to S', and these meet S again in Q and R, the six lines BC, CA, AB, PQ, PR and QR all touch a conic (above, Ex. 7); and this is the conic S', because S' touches five of these lines. In other words, an infinite number of triads of points P, Q, R can be found on S, for which the joins QR, RP, PQ touch S'. Take now one of these triads, P, Q, R. By what is proved above, in Ex. 21, there is a conic, say Ω, for which both A, B, C and P, Q, R are self-polar triads. Consider the conic which is the polar reciprocal of S

in regard to Ω. This touches the six lines BC, CA, AB, QR, RP, PQ (which are the polars in regard to Ω, of the points of the triads); it is therefore the conic S'. It follows from this that, if P', Q', R' be another triad of points of S for which $Q'R', R'P', P'Q'$ touch S', then P', Q', R' must be a self-polar triad in regard to Ω.

To prove this, let the polar of P', in regard to Ω, meet S in Q_1 and R_1; then, if the polar of Q_1, in regard to Ω (which polar passes through P'), meet $Q_1 R_1$ in R_2, a conic contains A, B, C, P', Q_1, R_2 (Ex. 12, above), which is therefore S; thus R_2 coincides with R_1, and P', Q_1, R_1 is a self-polar triad in regard to Ω. Therefore, P', Q_1, R_1 are points of S, the lines $P'Q_1, P'R_1, Q_1 R_1$ are tangents of the polar reciprocal, S'. Hence Q_1, R_1 are the same as Q', R'. We have therefore an infinite number of triads of points of S, of which the joins touch S', which are self-polar in regard to Ω.

There are, in fact, as has been remarked, three other conics beside Ω for which S and S' are polar reciprocals. These have not the same property in regard to the triads P, Q, R. And, in general, when S and S' are polar reciprocals in regard to a conic Ω, it is not the case that triads of points of S are self-polar in regard to Ω. The matter arises again for consideration in Chap. III (Ex. 24).

A conic S which contains triads of points A, B, C for which the joins BC, CA, AB all touch another conic Σ, may be spoken of as triangularly circumscribed to Σ. Let $S_1, S_2, \ldots$ be a set of conics all triangularly circumscribed to Σ. Let Ω_1 be the conic by which S_1 is the polar reciprocal of Σ, the triads spoken of being self-polar in regard to Ω_1; let Ω_2 similarly arise from S_2 and Σ, and so on. The set of conics $\Omega_1, \Omega_2, \ldots$ are then all outpolar to Σ.

Similarly, if we have a set of conics $\Sigma_1, \Sigma_2, \ldots$, all triangularly inscribed to a conic S, we can find a set of conics $\Omega_1, \Omega_2, \ldots$ all inpolar to S.

Ex. 25. Gaskin's theorem for conics through two fixed points outpolar to a given conic. Let Ω be a given conic, and I, J be two given points. We have defined (Ex. 3, above) the director conic of Ω in regard to I and J. We have considered two conics such that the tangents at their four common points intersect, in fours, in two points (Ex. 19, above). We have defined a conic as outpolar to Ω when there are triads of points of the conic which are self-polar in regard to Ω (Ex. 11). We now prove that all conics outpolar to Ω passing through the points I, J, are in the special relation with the director conic which was considered in Ex. 19. Such an outpolar conic is obtained by taking any triad which is self-polar in regard to Ω, and drawing the conic through I, J which contains the points of this triad. We have proved (Ex. 11) that any conic through the common points of two conics which are both outpolar to Ω is also outpolar to Ω; in particular any one of the line pairs through these common points consists of two lines conjugate to one another in regard to Ω. Thus, if any two conics outpolar to Ω be described through the points I, J, and X, Y be the remaining intersections of these, then the line XY passes through the pole of IJ in regard to Ω; this pole we denote by O. We denote, also, by A, the intersection of the lines IY and JX; as these lines are conjugate in regard to Ω, the point A is on the director conic of Ω in regard to I and J. Similarly, the point B, in which the lines IX, JY intersect, is on the director conic. This director conic passes through I and J, and, as we have remarked (Ex. 3), the tangents at I, J, of this conic, meet in the pole, O, of the line IJ taken in regard to the conic Ω.

Now draw any line through O, and consider the involution of pairs of points determined thereon by all conics through the four points I, J, X, Y. We prove that the double points of this involution are the points, say L and L', in which this line meets the director conic. One pair of this involution consists of the point O and the point, T, in which the line meets IJ. As O is in the pole of IJ in regard to the director conic, the points O and T are harmonic conjugates in regard to L and L'. Let the line, drawn through O, meet the lines IY and JX, respectively, in M and M'. As these lines join

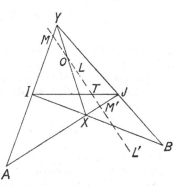

the point A, of the director conic, to the points I, J of this, and O is the pole of IJ in regard to this conic, it follows (Ex. 5, above) that M, M' are conjugate points in regard to the director conic, and are, therefore, harmonic conjugates in regard to L and L'. The points M, M' are, however, another pair of points determined, on the line OLL', by a conic through the four points I, J, X, Y; this pair, together with C and T, determine the involution. It is, therefore, shewn that L and L' are the double points of this involution.

Hence, the conic drawn through I, J, X, Y to contain L has, for its tangent at L, the line OLL'; as this passes through the pole, O, of IJ in regard to the director conic, the conic $IJXYL$ is related to the director conic as were the conics in Ex. 19. The conic $IJXYL$ was obtained as passing through the intersections of any two conics through I, J which are outpolar to Ω, and through any point L of the director conic. It is, therefore, in effect, any conic through I, J outpolar to Ω. And thus we have proved what was stated.

The result was given by Gaskin (*Geometrical construction of a conic section subject to five conditions*, Cambridge, 1852, p. 33, Prop. 15).

Ex. 26. *Two related ranges upon a conic obtained by composition of two involutions.* Referring for a moment to the symbolic representation, in order to explain the meaning of the result to which we now pass, we have remarked (Preliminary, above, Ex. 2), that two related ranges of points, $O + xU$, $O + yU$, are given by one of the two relations

$$\frac{y-\beta}{y-\alpha} = \sigma \frac{x-\beta}{x-\alpha}, \qquad \frac{1}{y-\alpha} = \frac{1}{x-\alpha} + \frac{1}{\lambda}.$$

Considering the former, if we take μ arbitrary, and take z so that

$$\frac{x-\beta}{x-\alpha} \cdot \frac{z-\beta}{z-\alpha} = \mu,$$

then we have

$$\frac{y-\beta}{y-\alpha} \cdot \frac{z-\beta}{z-\alpha} = \sigma\mu.$$

The first of these expresses that the points, $O + xU$, $O + zU$, are a pair of a certain involution; the second expresses that $O + zU$, $O + yU$ are a pair of another involution.

Similarly the latter equation is obtained by the composition of the two,

$$\frac{1}{x-\alpha} + \frac{1}{z-\alpha} = \frac{1}{\nu}, \qquad \frac{1}{y-\alpha} + \frac{1}{z-\alpha} = \frac{1}{\nu} + \frac{1}{\lambda},$$

wherein ν is arbitrary; and each of these connects a pair of points belonging to an involution.

Now consider two related ranges upon a conic, the general point of one range being P, the point of the other range corresponding to this being P'. Suppose that the pairs, P, P', do not belong to an involution; and, in the first instance, that the common corresponding points of the two ranges are not coincident. It is to be shewn that we can, in an infinite number of ways, find a range of points P_1, upon the conic, such that the pairs, P, P_1, form an involution, and the pairs P_1, P' also form an involution. Thus we may, in an infinite number of ways, regard a correspondence of related ranges upon a conic, as compounded of two involutions.

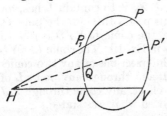

Let U, V be the common corresponding points of the two related ranges, and, on the line UV, take an arbitrary point, H. Let the lines joining H to any two corresponding points, P and P', of the two ranges, meet the conic again, respectively, in P_1 and Q. Then, as P varies, since the lines P_1P, QP', UV meet in a point, the pairs P_1, P; Q, P'; U, V belong to an involution. Therefore, the range P, Q, U, V is related to the range P_1, P', V, U, and is, hence, related to the range P', P_1, U, V. Wherefore, in the two original related ranges, to the position Q of the point P, corresponds the position P_1 of the point P'.

The ranges $(P_1), (P')$ are related; for, as P_1, P are pairs of an involution, the range (P_1) is related to (P); and this, by the original hypothesis, is related to (P'). And, when P_1 is at P', the point P, by the construction, is at Q. To this position of P, however, we have shewn, corresponds the position P_1 for P'. Namely, in the two related ranges $(P_1), (P')$, when P_1 is at P', the point P' is at P_1. Thus (Preliminary, above, p. 2), the two related ranges $(P_1), (P')$, together, form an involution. Therefore, the line P_1P' passes through a fixed point, say, K. This point is on UV; for when P is at V, then P_1 is at U and P' is at V.

We can, therefore, pass from the range (P) to the range (P'), by first passing to (P_1), where P, P_1 are pairs of an involution, the lines P_1P meeting UV in a fixed point, H, which may be taken arbitrarily, and, then, passing from (P_1) to (P'), the pairs P_1P' forming another involution, the lines P_1P' meeting UV in another fixed point, K, whose position depends upon that taken for H. Clearly K may be taken arbitrarily, if preferred, and H determined accordingly.

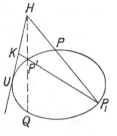

The theorem, and the proof, are, essentially, unchanged when the points U, V coincide. The line UV is then replaced by the tangent at U.

The proposition will be of use, later, in discussing the composition of two general screw motions. For this, and other purposes, it is desirable to state it in terms only of points which lie on the conic, without H and K. If L and L' be the points of contact with the conic of the two tangents which can be drawn from H, then, thinking first of U and V as distinct, the points L and L', as points of the conic, are harmonic conjugates in regard to U and V; for the lines joining L to H and L' are harmonic conjugates in regard to the lines LU, LV Similarly, L and L', are harmonic conjugates in regard to P_1 and P. Thus, in place of taking H arbitrarily on UV, we may take any two points L, L', of the conic, which are harmonically conjugate, on the conic, in regard to U and V. Then P_1 will be obtained from P by the fact that, as points of the conic, P_1 and P are harmonic conjugates in regard to L and L'. There will then be two other fixed points of the conic, say, M and M', also harmonic conjugates in regard to U and V, in regard to which P_1 and P' are harmonic conjugates. The points M, M' are the points of contact of the tangents from K. When U and V coincide, in U, we are, similarly, to take a quite arbitrary point, L, of the conic, and determine P_1 as the harmonic conjugate of P in regard to U and L; there will then be another fixed point, M, of the conic, such that P_1 and P' are harmonic conjugates in regard to U and M.

Ex. 27. *Three triads of points of a conic are all in perspective with another triad of points of the conic.* If A, B, C; A', B', C', and A'', B'', C'', be any three triads of points of a conic, we can find another triad of points of the conic, A_1, B_1, C_1, which is in perspective with each of the three given triads, from three properly chosen centres. For if we regard A, B, C and A', B', C' as determining two related ranges upon the conic, wherein the points A, B, C, of the first range, correspond, respectively, to the points A', B', C', of the second, these ranges will have two common corresponding points on the conic, whose joining line is the axis of relation of the two ranges, containing all such cross intersections as that of AB' and $A'B$, of BC' and $B'C$, etc. By taking the three triads in pairs, we have, thus, three such axes of relation, whose intersections are three in number. These intersections are the centres of perspective spoken of. This follows from Ex. 26; the range (A, B, C) is related to the range (A', B', C') by means of two involutions, one of these, with the pairs A, A_1; B, B_1; C, C_1, with the use of one of the centres of perspective, the other, with the pairs A_1, A'; B_1, B'; C_1, C', with the use of another of the centres of perspective. And so on.

Though out of logical order, it may be as well to remark that this result is identical with one that, in the language of metrical geometry, which we consider later (Chap. v, below), is expressed by saying that three congruent figures in a plane may be regarded as images, by reflexion, of the same figure, in three suitably taken mirrors.

Ex. 28. *The envelope of the line joining two corresponding points of two related ranges on a conic.* We have seen that the joins of pairs of points of an involution on a conic, all pass through a point. Also, that the joins of corresponding points, of two related ranges of points, on two lines, are tangents of a conic. In generalisation of these we now prove that the joins of two corresponding points, of two related ranges of points on a conic, are tangents of another conic, which touches the original, at the two (generally distinct) points which are the common corresponding points of the two ranges.

Let the corresponding points be P, P'; let O be an arbitrary point of the plane. There are two of the lines PP' which pass through O. For, let the relation between the ranges $(P), (P')$ be given, as in Ex. 26, by means of lines, PP_1, passing through a fixed point, H, and lines, P_1P', passing through a fixed point, K. Let OP meet the conic again in Q. The pairs (Q, P), (P, P_1), (P_1, P') are then pairs of three involutions; from this it can be shewn that the ranges (Q), (P') are related. These two related ranges will have two common corresponding points, in general distinct; but when Q coincides with P', there is a line PP' passing through O.

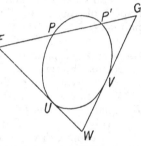

Assuming this result, let U, V be the common corresponding points of the ranges (P), (P'); and let F be any point of the tangent at U. There will then pass through F two lines PP', of which, however, FU is one. The other will meet the tangent at V in a definite point, G. Similarly, to an arbitrary position of G, on the tangent at V, there corresponds a single definite position of F, on the tangent at U. Hence, by an abbreviated argument, spoken of in the Preliminary remarks (above, p. 8), we infer that the ranges (F), (G) are related; and hence, that the lines PP' touch a conic; this conic touches the tangent at U at the point where F lies when G is at the intersection, W, of the tangents at U, V, that is at U. It similarly touches VW at V.

It is easy to state the corresponding result when U and V coincide.

Ex. 29. The tangents from three points, A, B, C, to a conic, meet BC, CA, AB in six points of a conic. Let the tangents from three

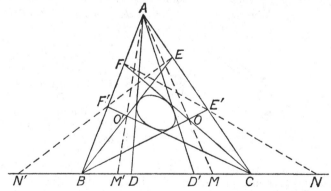

points, A, B, C to an arbitrary conic Ω, meet BC, CA, AB, respectively in D, D', in E, E', and in F, F'. These six points then lie on another conic.

Let BE', CF meet in O, and AO meet BC in M; let BE, CF' meet in O', and AO' meet BC in M'. Let $E'F$ meet BC in N, and EF' meet BC in N'.

We have shewn that the tangents from an arbitrary point to the conics touching four lines are a pencil in involution; and of such conics there are three point pairs, through each of which the four lines pass. For the four lines constituted by the tangents to the conic Ω from B and C, two of these point pairs are B, C and O, O'. Hence we see that the pairs D, D'; M, M'; B, C, lying on tangents from A, are in involution.

The point N is, however, the harmonic conjugate of M in regard to B and C, and the point N' is the harmonic conjugate of M', in respect of the same. Thus (see Preliminary, above, Ex. 3), the points N, N' also belong to the involution spoken of.

Now the conics drawn through the four points E, E', F, F' meet the line BC in pairs of points belonging to an involution; of this involution, however, B, C and N, N' are two pairs; so that this is the same involution as the former.

Wherefore, the conic through E, E', F, F' which is drawn through D, equally contains D'. And this is what we desired to prove.

If this conic be called Σ, the reader may proceed to prove that there is a conic, through the intersections of Ω and Σ, which touches BC, CA, AB.

A particular case of the result proved, arises when the conic Ω is a point pair. Of this again a particular case is that, if AD, BE, CF meet in a point, then a conic exists touching BC, CA, AB, respectively, at D, E, F. And the theorems lead, by duality, to other theorems worth notice. (See also Chap. III, Ex. 14.)

Ex. 30. *Two tangents of one conic meet two tangents of another in four points lying on a conic through the common points of the two conics, if the four points of contact are in line.* Let a line joining

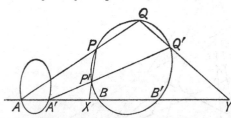

two points, A and A', of one conic, Φ, meet another conic, Ω, in B and B'. Let any line through A meet Ω in P, Q, and any line through A' meet Ω in P', Q'. Let PP', QQ' meet the line AA', respectively, in X and Y.

The pairs of points A, A'; B, B'; X, Y then belong to the involution determined on the line AA' by all conics through the four points P, Q, P', Q'. And as this involution is determined by the first two pairs, it is the same as the involution determined on the line AA' by all conics through the common points of the conics Φ, Ω. To this latter involution, therefore, the pair X, Y belong.

By supposing A' to coincide with A, and P' with P, and Q' with Q, we reach the result that, if any line, drawn through a point, A, of the conic Φ, meet the conic Ω in P and Q, and the tangents at P, Q meet the tangent at A in X and Y, then these last are a pair of the involution determined, on the tangent at A, by the conics which pass through the four common points of Φ and Ω.

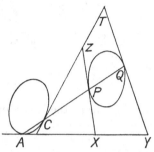

Now let the line APQ meet the conic Φ again in C, and let the tangents at P and Q meet the tangent at C in Z and T. Then, by parity of reasoning, taking P on Ω instead of A on Φ, and A, C on Φ instead of P, Q on Ω, we infer that the conic drawn through the intersections of Φ and Ω to contain X also contains Z; from what was said, this conic contains Y. By a similar argument, the same conic contains T.

This is the result enunciated, the points A, C, P, Q being in line.

Dually, if from an arbitrary point the two pairs of tangents be drawn to two conics, the four joins, of the points of contact on one conic, to the points of contact on the other conic, are all tangents of a conic which touches the four common tangents of the two conics.

From the combination of these theorems many interesting results can be deduced. It will be proved in Chap. III, that, if the tangents of the conics Φ, Ω, respectively at A and P, meet in X, and the conic, through the common points of Φ, Ω, which contains X, be S, while that, touching the common tangents of Φ and Ω, which touches AP, be Σ, then, to another position of X on S, there corresponds, by the same construction, another tangent, such as AP, of Σ.

Ex. 31. Poncelet's theorem, for sets of points A, B, C, ... of one conic, of which the joins AB, BC, ... separately touch other fixed conics, all the conics having four points in common.

The reference is to Poncelet's *Propriétés Projectives*, Paris, 1865, t. I, p. 316.

Let A, A', B, B' be four points of a conic, S, such that the lines AB, $A'B'$ touch another conic, S', respectively, in P and P'. Let PP' meet AA, BB', respectively, in H and K. We shew, first, that a conic can be drawn, through the common points of S and S', to touch AA' in H, and BB' in K.

For consider the four points consisting of P repeated and P' repeated; and the involution determined on AA' by the conics through

these four points. Three pairs of this are (1), the points A, A', on
the conic S, (2), the points in which the conic S' meets AA', (3), the

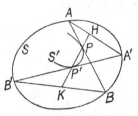

point H repeated. The pairs (1), (2) shew
that this is the same involution as that
determined on AA' by the conics through
the common points of S and S'. Of this
involution, then, H is a double point, and
is, thus, a point of contact with AA' of a
conic, say, T, through the common points
of S and S'. We prove, now, that this
conic, T, touches BB' at K. This we do by
shewing, from the preceding example, that, if T meet HK again
in K_1, the tangent of it, at K_1, passes through B and B'. By that
example, the conics S', T being met by a line in P, P' and H, K_1,
the tangents at P, P' must meet the tangents at H, K_1, in four
points lying on a conic passing through the common points of S'
and T, which are the common points of S and S'. These four points
are A, A', and the points where the tangent at K_1 meets AB and
$A'B'$; the two former points identify the conic through them as S.
This proves that the tangent at K_1 passes through B, B'; so that
K_1 is K.

It follows, of course, in the same way, that if PP' meet AB' and

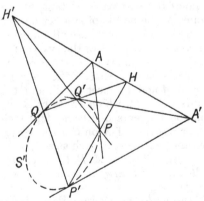

$A'B$ in L and M, a conic
through the common points
of S and S' touches these lines
in L and M.

If the other tangent from
A to S' touch this in Q, and
the other tangent from A' to
S' touch this in Q', it follows,
similarly, that PQ' meets AA'
in a point, H', which is also a
point of contact with AA' of
a conic through the common
points of S and S'. Thus, H'
is the other double point of
the involution spoken of, and
is the harmonic conjugate of H in regard to A and A'. But, then,
by the same argument, $Q'Q$ meets AA' in the harmonic conjugate
of H', that is, in H; and, so, $P'Q$ passes through H'.

Now take a further conic, S'', passing through the common points
of S and S', or, say, for brevity, a conic of the system (S, S'). Let
either one of the tangents drawn from B to S'' meet S again in C,
touching S'' in R. It follows, then, from what has just been said,
that it is possible to pair with this such an one of the two tangents

from B' to S'', touching S'' in R', that the line RR' shall pass through K; this point, K, is one of the two points of possible contact, with BB', of conics of the system (S, S'). If the appropriate tangent meet S again in C', the line RR' will meet CC' in a point, J, in which the conic (S, S'), which touches BB' at K, touches CC'.

There are now two pairs of points of S, the pairs A, A' and C, C', such that the lines AA' and CC' both touch the same conic of the system (S, S'), respectively at H and J. Call this conic Σ. By applying to the lines AA', CC', and the conics S and Σ, the argument we originally applied to the lines AB, $A'B'$, and the conics S and S', we can then infer that the lines AC, $A'C'$ are both touched by a further conic, S''', of the system (S, S'), the points of contact being on the line JH.

It is thus possible to have an infinite number of triads A, B, C; A', B', C'; ... of points of S, for which the joins $BC, B'C', ...$; $CA, C'A', ...$; $AB, A'B', ...$; touch, respectively, three fixed conics S', S'', S''', of the system (S, S'). The argument can clearly be extended to the case when the points of a set are not triads, but of any greater number.

A particular case of the foregoing arises when, with four points, A, A', B, B', of the conic S, of which $AB, A'B'$ meet in a point O, we have, instead of the proper conic S', a pair of lines which intersect in O. There are then no distinct points of contact, P, P', with which to determine the points H, K of AA' and BB'. But, a point H can be determined on AA', as one of the two which are harmonic, both in regard to A, A', and in regard to the two points in which AA' is met by the lines of the line pair, S'. The line HO then meets BB' in a point, K, which is one of the two, which are harmonic conjugates, both in regard to BB', and in regard to the two points in which BB' is met by the lines of the line pair, S'. A conic of the system (S, S') can then be drawn to touch AA' at H, and to touch BB' at K.

If, in the preceding theory, we start with two proper conics, S and S', a line pair, of the system (S, S'), can intersect only in a point, O, which is one of the three points of the common self-polar triad of the system. Such a line pair can then replace S'' or S'''.

In particular, we may have triads, A, B, C, of points of S, such that AB, BC touch two proper conics, S', S'' of the system (S, S'), respectively in P and Q, say, while CA passes through one of the three points, O, of the self-polar triad of (S, S'). It will appear (in Chap. III) that, in this case, the line PQ is a tangent of a fixed conic, which touches the common tangent lines of S' and S''; and that the line pair of (S, S') which meet in O are harmonic conjugates in regard to the lines OP, OQ.

A very particular case is the intersecting simple result that triads *A, B, C* exist on a conic, of which the joins *BC, CA, AB* pass, respectively, each through one of the points of any given triad which is self-polar in regard to the conic.

Ex. 32. Construction of a fourth tangent, with its point of contact, of a conic touching three lines at given points.

The present example, which, in effect, collects various results already given, is introduced in view of the following example.

Respectively on the joins *BC, CA, AB*, of three points *A, B, C*, let *D, E, F* be points such that the lines *AD, BE, CF* meet in a

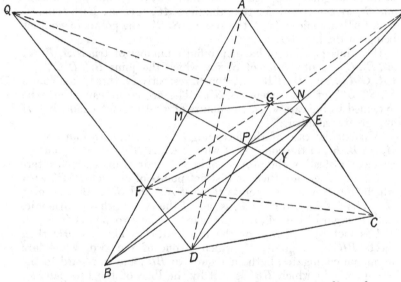

point. Let the lines *DE, DF*, respectively, meet any line, drawn through *A*, in *R* and *Q*. Let *CQ, BR* meet *AB, AC*, respectively, in *M* and *N*, the lines *CQ, BR* intersecting in *P*. We prove that the lines *QR, MN, BC* meet in a point; and that the lines *MN, DP, EQ, FR*, meet in a point; while *P* is on *EF*. Also that *MN* is a tangent of the conic which can be drawn to touch *BC, CA, AB*, respectively, at *D, E, F*, whose point of contact is on *DP*, as well as on *EQ* and *FR*.

In fact *FE* meets *BC* in the point harmonic to *D* with respect to *B* and *C*; thus *F, E* are harmonic with respect to the points in which *FE* is met by *DA* and *BC*. Therefore, *Q, R* are harmonic with respect to *A*, and the point where *QR* is met by *BC*. This shews that *QB, RC* meet on *AP*, so that the triads *QMB, RNC* are in perspective, the lines *QR, MN, BC* meeting in a point. That *P* lies on *FE* is a direct application of Pappus' theorem for the two triads Q, A, R and B, D, C. That QE, FR meet on MN is, then, also a consequence of Pappus' theorem, applied to the two triads Q, A, R and F, P, E. To shew that DP passes through the point, say, G, where MN is met by EQ or FR, it is necessary to prove that E, F are harmonic in regard to P, and the point where QR meets EF; or, that AP, AR are harmonic in regard to AE, AF, which follows because QB, RC meet on AP. Finally, that the conic which (Ex. 29) can be drawn to touch BC, CA, AB at D, E, F, respectively, touches MN, on DP, follows from Ex. 14, above; or can be deduced, independently, by considering, with fixed points D, E, F, various lines QAR, giving rise to related ranges (M), (N), on AB, AC. In regard to this conic, P, Q, R will be a self-polar triad.

It is clear that B and N are harmonic conjugates in regard to P and R; and, further, if QP meet DE in Y, that D and E are harmonic conjugates in regard to Y and R.

Ex. 33. Hamilton's extension of Feuerbach's theorem. (See, also, p. 117.)

With the results of the previous example in mind, consider two conics. One of these, Ω, is to be any conic having P, Q, R for a self-polar triad. The other, Σ, is to be the conic through D, E, F which meets the line MN in the points in which it is met by Ω. We prove that these conics meet further on the line, which we call l, which contains the three intersections (BC, EF), (CA, FD), (AB, DE). This is the line which, for the conic touching BC, CA, AB at D, E, F, we have called the Hessian line of D, E, F (Ex. 9). In fact, the conic Ω meets the line DR in points harmonic in regard to Y and R; the conic Σ also meets this line in points, D and E, harmonic in regard to Y and R. Wherefore the common chord MN, of Ω and Σ, and the line joining the other two common points of these conics, meet DR in points harmonic in regard to Y and R. The point harmonic to the intersection of MN with DR, with respect to Y and R, is, however, we have seen, the point where MB meets DR. This is a point of the line, l, spoken of. Another point of the common chord of Ω and Σ, other than MN, is proved, similarly, to be on l, by considering the line DQ, in place of DR.

This is a slight extension of a result ascribed to Sir W. R. Hamilton (Salmon's *Conics*, 1879, p. 313). For what follows we take for Ω any conic touching BC, CA, AB and touching MN, say, at K; this will have P, Q, R for a self-polar triad; then Σ will be the conic through D, E, F which touches MN at K. The result is that the join, of the two remaining intersections of these two conics, is the line l.

Conversely, let A, B, C be any points, and l any line. Let D, on BC, be the harmonic, in regard to B and C, of the point where l meets BC; and make a similar construction for E and F. Let Ω be any conic touching BC, CA, AB. Let its intersections with l be I and J. Then the conic Σ, defined as that containing D, E, F, I, J, touches Ω. This is the theorem desired.

The point of contact of Σ with Ω may be defined as the point of contact with Ω of the fourth common tangent of Ω with the conic which touches BC, CA, AB at D, E, F, respectively. But it may be defined without this conic. For, if Ω touch BC, CA, AB respectively at D', E', F', the point P (one of the three points of the common self-polar triad of Ω and the conic which touches BC, CA, AB at D, E, F) is the intersection of $E'F'$ and EF, by what we have seen. Hence the points M, N can be constructed, as in the preceding example, by the lines CP, BP (as also the points Q and R). The point, K, required, is then the intersection of $D'P$ with MN (which also lies on $E'Q$ and $F'R$). Incidentally, given the points of contact of two conics with BC, CA, AB, we thus construct their remaining common tangent, and its points of contact with the two conics.

When the points A, B, C, I, J are given, we can (Ex. 10, above) find a point, T, such that each of the line pairs, AT, BC; BT, CA; CT, AB, meet the line IJ in points which are harmonic in regard to I and J. The conic Σ will then be (Ex. 18) the locus of the poles of the line IJ in regard to all conics through A, B, C, T. Let then A, B, C, T be four arbitrary points, and let I, J be any two points conjugate to one another in regard to all conics through A, B, C, T. Four conics, such as Ω, can (Ex. 17) be constructed, to pass through I, J, and touch the joins of three of the four points A, B, C, T. All the sixteen conics so obtained are touched by the conic which is the locus of the poles of IJ in regard to the conics through A, B, C, T.

Feuerbach's theorem in its original form will be referred to in the next Chapters. (See pp. 75, 89, 113, 122.)

Ex. 34. Theorem in regard to the remaining common tangents of three conics which touch three given lines.

If three conics, S_1, S_2, S_3, all touch the lines BC, CA, AB, the points of contact being, in the case of S_1, respectively D_1, E_1, F_1, with a similar notation for the others, and, if the remaining common tangent of S_2 and S_3 meet BC in T_1, with a similar notation for the other pairs, then the pairs T_1, D_1; T_2, D_2; T_3, D_3 are in involution.

For, by what has been said, in the two preceding examples, the remaining common tangent of S_2 and S_3 meets AB, AC, respectively, in M_1 and N_1, where N_1 and M_1 are, respectively, on the lines joining B and C to the point, P_1, in which $F_2 E_2$ intersects $F_3 E_3$.

Thus the point T_1, where $M_1 N_1$ meets BC, is the harmonic conjugate, in regard to B and C, of the point in which AP_1 meets BC. The point D_1 is also the harmonic conjugate in regard to B and C, of the point where F_1E_1 meets BC. This line F_1E_1 is the same as the line P_2P_3. It is a fundamental property of the joins of four points, that the joins of A, P_1, P_2, P_3 meet BC in pairs of points in involution. As these points are the harmonic conjugates, in regard to B and C, of the pairs T_1, D_1; T_2, D_2; T_3, D_3, it follows that these are also in involution. (Preliminary, above, p. 7, Ex. 4.)

The dual theorem is that, if three conics be drawn through three points A, B, C, and we draw the three lines from A to the remaining intersections of the pairs of these conics, and pair these, properly, with the tangents at A of the three conics, there is obtained a pencil in involution.

A more general proposition is that, if any four conics be drawn through the three points A, B, C, say S_1, S_2, S_3, S_4, then the three pairs of lines through A constituted by (1) the common chords through A of S_2, S_3 and of S_1, S_4; (2) the common chords through A of S_3, S_1 and of S_2, S_4; (3) the common chords through A of S_1, S_2 and of S_3, S_4, are in involution. This may be proved as above, or also, by the transformation of Ex. 16 preceding. It depends on the fact that the pairs of lines from an arbitrary point, to the opposite pairs of intersections of four lines in a plane, belong to an involution. (Preliminary, p. 4, above.) When the fourth conic breaks up into two lines, of which one passes through A, we obtain the theorem referred to in regard to three conics through A, B, C.

Ex. 35. *The four conics through two points of which each contains three of four other given points.*

If the two given points be O, Q, and the four given points be denoted by 1, 2, 3, 4, and the conic which contains $O, Q, 2, 3, 4$ be called S_1, its tangent at O being called t_1, with a similar notation for three other conics, then the four pairs of lines $(O1, t_1), (O2, t_2), (O3, t_3), (O4, t_4)$ are in involution. For the conics S_1, S_2, S_3 have in common the three points $O, Q, 4$, while S_2, S_3 have also the point 1 in common, S_3 and S_1 the point 2, and S_1, S_2 the point 3. The pairs of lines $(O1, t_1), (O2, t_2), (O3, t_3)$ are, thence, in involution, by the preceding example. Again the conics S_2, S_3, S_4 have in common the points $O, Q, 1$; the residual intersections of the pairs of these are, for S_3, S_4, the point 2; for S_4, S_2, the point 3; and, for S_2, S_3, the point 4. Wherefore $(O2, t_2), (O3, t_3), (O4, t_4)$ are in involution; this is, therefore, the same involution as before.

Ex. 36. *The polars of any point in regard to the six conics which can be drawn through the sets of five, from six arbitrary points of a plane, touch a conic.*

This result was suggested by Mr H. M. Taylor, Trinity College, Cambridge. (See W. W. Taylor, *Messenger of Mathematics*, xxxvi, 1907, p. 113; and particularly, E. J. Nanson, *Mess. of Math.* xlvi, 1917, p. 183.)

Let the six given points be denoted by 1, 2, ..., 5, 6, the other point being O, the conic containing the five points 2, 3, 4, 5, 6 be S_1, etc., the polar of O in regard to S_1 being p_1, etc. Any two of the four conics have four points in common, and a further, auxiliary, conic can be drawn through O and these four points. As the polars of O, in regard to conics through four points, meet in a point (Ex. 15, above), the polars of O in regard to the two conics meet on the tangent, at O, of the auxiliary conic spoken of.

Consider then the intersections of p_5 with p_1, p_2, p_3, p_4. These lie, by what has been said, on the tangents, at O, respectively of the four conics containing the points

$$O, 2, 3, 4, 6; \quad O, 1, 3, 4, 6; \quad O, 1, 2, 4, 6; \quad O, 1, 2, 3, 6.$$

These however are conics, through the points O and 6, containing, respectively, the various sets of three points which can be chosen from the four 1, 2, 3, 4. By the result of the preceding example, it follows that the range of four points in which p_5 is met by p_1, p_2, p_3, p_4 (these points lying on the tangents at O of the four conics just named) is related to the pencil of lines joining O to the points 1, 2, 3, 4. The range in which p_6 is met by p_1, p_2, p_3, p_4 is, similarly, related to this pencil. Whence the six polar lines touch a conic.

The above interesting proof of Taylor's theorem is derived from that given by Mr S. G. Soal, of the East London College, in J. L. S. Hatton's *Principles of Projective Geometry*, Cambridge, 1913, p. 299.

From this demonstration we can state the theorem: *Given six arbitrary points of a plane, there exists, corresponding to any further arbitrary point, O, of the plane, a conic, and, upon this conic, six points subtending, at any point of the conic, a pencil which is related to that of the lines which join O to the six given arbitrary points.* For the points of contact of p_1, p_2, p_3 p_4 with the conic are, as a range on the conic, related to the range determined by these tangents on the tangent p_5.

We shall be further concerned, in a later volume, with the conics passing through the sets of five of six given arbitrary points.

CHAPTER II

PROPERTIES RELATIVE TO TWO POINTS OF REFERENCE

Introductory. The multiplicity of metrical results often found in books on Geometry are in fact relations of the figure which is considered, either to two points of reference, regarded as given, or, more generally, to a conic of reference, regarded as given. This is one of the striking discoveries, essentially due to Poncelet, of the early part of the nineteenth century. The fact is generally proved by beginning, on the basis of Euclid's geometry, with the metrical relations, and gradually shewing how these can be summarised under more general points of view. We adopt the converse procedure. Without assuming the metrical basis, and as consequences of the foundations which have been explained, we obtain a number of results which are easily recognised as containing those usually developed as consequences of the notion of distance. Finally, from this and Chapter v, it should emerge quite clearly that distance is only a special way of using the algebraic symbols; and that it is an arbitrary limitation to regard it as fundamental.

For the sake of clearness we shall employ the terms, such as *perpendicular, circle, rectangular hyperbola*, and so on, which are current in metrical geometry. They have here, however, a more general meaning, depending on the choice of the absolute points of reference.

Middle points, Perpendicular lines, Centroid, Orthocentre. Suppose that two points of reference are given; they may be called the Absolute points, and will often be denoted by I and J.

If any line, AB, meet the line IJ in K, the point, say C, which is the harmonic conjugate of K in regard to A and B, may be called the *middle point* of AB, in regard to I and J. Any two lines whose intersection is on the line IJ may be said to be parallel to one another, in regard to I and J. Any two lines which meet the line IJ in two points which are harmonic conjugates of one another, in regard to I and J, may be said to be *at right angles*, or, to be *perpendicular*, in regard to I and J. With these definitions, all the theorems of incidence involving these terms remain true. Thus, for instance, two lines, which are both parallel to a third line, are parallel to one another; all lines which are at right angles to a line are parallel to one another; if a series of parallel lines be drawn, meeting any two lines, say a and b, the former in A and the latter in *B*, the middle point of *AB* lies on a definite line through the intersection of *a* and *b*.

Or, passing to somewhat less obvious results, the ordinary constructions for the centroid, and for the orthocentre, of three given points *A*, *B*, *C*, can be carried out, these words being defined as we now explain.

For the *centroid*, this is defined as the intersection of the lines joining *A*, *B*, *C*, respectively, to the middle points of *BC*, *CA*, *AB*.

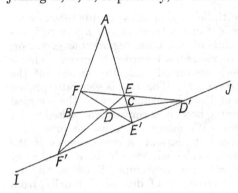

If the line *IJ* be met by *BC*, *CA*, *AB*, respectively, in *D'*, *E'*, *F'*, and *D* be the harmonic conjugate of *D'* in regard to *B*, *C*, while *E* is similarly defined from *E'* by means of *C* and *A*, and *F* from *F'* by *A* and *B*, we have only to shew that *AD*, *BE*, *CF* meet in a point. In fact, it is clear, from the construction, that *E*, *F*, *D'* are in line; as, likewise, are *F*, *D*, *E'* and *D*, *E*, *F'*; and, hence, that the line, joining *A* to the intersection of *BE* and *CF*, passes through *D*.

For the *orthocentre*, we are to shew that the lines, drawn through *A*, *B*, *C*, respectively perpendicular to *BC*, *CA*, *AB*, meet in a point. As before, let *D'*, *E'*, *F'* be the points in which *IJ* is met by *BC*, *CA*, *AB*, respectively; and let *U*, *V*, *W* be the harmonic conjugates, in regard to *I* and *J*, respectively of *D'*, *E'*, *F'*. We are to shew that *AU*, *BV*, *CW* meet in a point.

If *AU*, *CW* meet in *P*, the involution determined on the line *IJ*

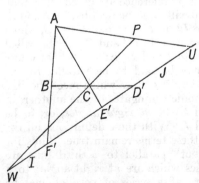

by the joins of *A*, *B*, *C*, *P* (Preliminary, above, p. 4), is determined by the two pairs *D'*, *U* and *F'*, *W*. In regard to these pairs, *I*, *J* are harmonic conjugates; these are then the double points of the involution, and thus *AC*, *BP* meet the line *IJ* in points harmonic in regard thereto. Thus *BV* passes through *P*. Each of *A*, *B*, *C*, *P* is then the orthocentre of the other three in regard to *I* and *J*.

Considering I, J as a degenerate conic, regarded as an envelope (above, p. 24), the theorem of the orthocentre is evidently a particular case of the theorem (above, p. 31) that, if, of the line pairs containing four points, two pairs consist of lines which are conjugate in regard to a conic, so does the third pair.

Circles. Any conic which passes through the two absolute points of reference, I, J, will be spoken of as a circle in regard to I and J, or, simply, as a circle. By the *centre* of the circle will be meant the pole, in regard thereto, of the line IJ. As a conic is determined by five points, a circle is determined by three points, beside I and J.

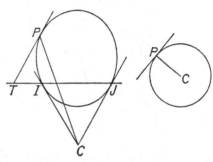

If the tangent, at a point P, of a circle, whose centre is C, meet the line IJ in T, this point, T, lying on the polar of C, and on the tangent at P, is the pole of the line PC. Thus PT, PC meet IJ in points which are harmonic in regard to I and J.

In other words, the tangent at any point of a circle is at right angles to the line joining this point to the centre of the circle.

Two circles will have two points of intersection, beside I and J. The tangents of one of the circles, at these two points, meet on the line joining the centres of the circles. For, if these two points be denoted by A and B, and D, E, F be the intersections, respectively, of IJ, AB, of IA, JB, and of IB, JA, the pole, in

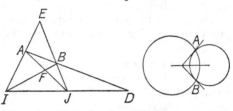

regard to either circle, of the point D, is the line EF (above, p. 23), and this line contains the intersection, both of the tangents at A, B, and of the tangents at I, J. This line, EF, is thus the line joining the centres of the two circles.

Suppose that, at one of the common points, A, of two circles, the tangents of the circles are at right angles to one another, in regard to I and J. As the tangent of a circle at any point is at right angles to the line joining this point to the centre, it follows that the tangent, at A, of either circle, passes through the centre of the other. Thence, as the tangents of either circle, at the points A, B, meet on the line joining the centres of the circles, it follows that the tangent at B, of either circle, passes through the centre of

the other circle. Wherefore the tangents at *B*, of the two circles,
are at right angles. Two such circles are said to *cut at right angles*,
or *orthogonally* ; they are such a pair of associated conics as were
considered before (Chap. I, Ex. 19, above, p. 42). It follows from
what was then said that, any line, drawn through the centre of
either of two circles which cut at right angles, meets the two circles,

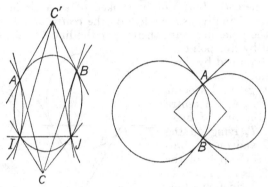

respectively, in two pairs of points which are harmonic in regard
to one another ; and, conversely, that if, on any line drawn through
the centre of a circle, there be taken two points which are harmonic
conjugates of one another in regard to the points in which the line
cuts the circle, then any circle drawn through these two points cuts
the given circle at right angles. It follows, also, that two points,
which are such that every circle, passing through them, cuts a
given circle at right angles, must be conjugate points in regard to
this circle.

Coaxial circles. The common chord, *AB*, of two circles, other
than *IJ*, may be called the *radical axis* of the two circles. Circles
passing through *A* and *B*, that is, conics through *A*, *B*, *I*, *J*, may
be said to be *coaxial*.

Of such conics, beside the lines *IJ*, *BA*, there are two which are
line pairs, the pair *IA*, *JB*, meeting, say, in *E*, and the pair *IB*, *JA*,
meeting, say, in *F*. The points *E*, *F* may be called the *limiting
points* of the coaxial circles through *A* and *B*. The line *EF* contains
the centres of all the coaxial circles ; and *E*, *F* are conjugate points
in regard to any one of these circles. The triad of points self-conjugate in regard to all the coaxial circles consists of the limiting
points *E*, *F*, together with the point, *D*, where the radical axis
intersects the line *IJ*. Thus, by what we have said above, all circles
through the limiting points, *E*, *F*, cut every circle of the coaxial
system at right angles ; these circles form a second coaxial system,
of which *A* and *B* are the limiting points, their centres being on

the radical axis, *AB*, of the first system. Conversely, a circle whose centre is at any point on the radical axis, *AB*, of the first system of circles, can be drawn, to cut all the circles of this system at right angles ; it passes through the limiting points.

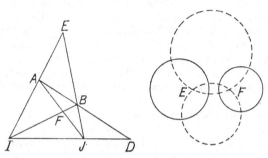

Inversion in regard to a circle. There is a process, which has been much employed, by which, when we are given a circle in a plane, we can pass from any point of the plane to another point. It will subsequently be found, in dealing with the relations of geometry in two and in three dimensions, that the process can be regarded from a very simple point of view. It can also be shewn that, in its effect, the process does not differ from the transformation explained in Chap. I, Ex. 16 (above, p. 38). We give, however, some provisional results obtained from the present point of view.

Two points, *P* and *P′*, are said to be inverses of one another in regard to a given circle when, (*a*) they are conjugate in regard to the circle, and, (*b*) their join passes through the centre of the circle. When this is so, it follows, by what has already been remarked in the preceding chapter (Ex. 5, above, p. 28), that the lines *IP*, *JP′* meet in a point, say, *L*, lying on the circle, and the lines *IP′*, *JP* also meet in a point of the circle, say, *M*. The relation of the points *P*, *P′* is evidently reciprocal. The definition can be held to include the inverse of a point in regard to a line, which, taken with the line *IJ*, may be regarded as a degenerate circle ; the inverse of *P* in regard to the line, say, *P′*, will be such that

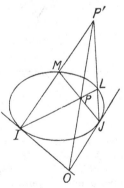

PP′ is at right angles to the line, and the mid-point of *PP′* is on the line.

When *P* describes any locus, *P′* will describe a corresponding locus. The most important case is that, when *P* describes any

circle, then P' also describes a circle. Of this a particular case
arises when P describes a circle which passes through the centre, O,
of the given circle by which we define the inversion; then the circle
described by P' degenerates into a line, taken with the line IJ.
Conversely, we may say that the inverse of a line is a circle passing
through the centre of the fixed circle of inversion. These statements
are easy to prove: When P describes any conic passing through I
and J, the pencils $I(P)$, $J(P)$ are related; these, however, give,
respectively, the ranges (L), (M), of the given fundamental conic,
which are, therefore, also related. Thus P', the intersection of JL
and IM, describes another conic passing through I and J, that is,
a circle. While, if P, on its locus, can take the position O, the
points L, M, take, then, respectively, the positions I and J; then
the ray IJ, of the pencil $I(P')$, corresponds to the ray JI, of the
pencil $J(P')$, and the locus of P' degenerates (above, p. 10) into
the line IJ and another line. Conversely, if P' describes a line, the
pencils, $I(P')$, $J(P')$, are related; so then, also, are the pencils
$I(P)$, $J(P)$; thus P describes a conic passing through I and J.
But, in particular, to the point P' which is on IJ, correspond
positions, of L and M, respectively at I and J; the corresponding
position of P is then at O. The locus of P is, thus, a circle passing
through the centre, O, of the circle of inversion.

Examples of inversion in regard to a given circle.
Ex. 1. Let the point P, of one circle, invert into the point P',

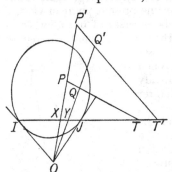

of the inverse circle; and let the
tangents of these circles, respectively at P and P', meet the line IJ,
in T and T'; let the line OPP'
meet IJ in X. Then X is a double
point of the involution determined
by the pairs I, J and T, T'. Consider, first, two points, P and Q;
and their inverses, respectively P'
and Q'. Let PP' and QQ' meet IJ,
respectively, in X and Y, while PQ
and $P'Q'$ meet IJ in T and T'.
Then, by what has been said above,
the conic drawn through I, J, P, Q, P' also contains Q'; or, the
points P, Q, P', Q' lie on a circle, cutting the circle of inversion
at right angles. Thus, the pairs X, Y; T, T'; I, J belong to the
involution, of points on IJ, determined by conics through P, Q,
P', Q'. When Q coincides with P, and Q' with P', this leads to
the result stated.

Ex. 2. The circles which are the inverses of two circles which
cut at right angles are two circles which also cut at right angles.

Let the tangents of two circles, which meet at P, and cut at right angles, intersect the line IJ, respectively, in T and U; let the tangents of the, respectively, inverse circles, at the inverse point P', meet IJ in T' and U'. Consider the involution determined by the two pairs of points, I, J and T, T'. The point, U, which is the harmonic conjugate of T in regard to I and J, and the point, V, which we may take, which is the har- monic conjugate of T' in regard to I and J, form a pair also belonging to this involution (above, p. 7). If the line $P'PO$ meet IJ in X, we have seen in the preceding example that the in- volution determined by I, J with X as a double point, contains the pairs T, T' and U, U'. By what we have now said,

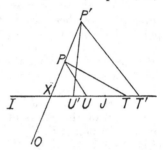

this same involution, as determined by I, J and T, T', contains the pair U, V. Thus V coincides with U', and the points T', U' are harmonic conjugates in regard to I, J. Therefore, the inverse circles cut at right angles at P'. It is easy to prove, also, that the inverses of two circles which touch one another are two circles, which also touch one another.

Ex. 3. Suppose we are given a circle and two points which are inverse to one another in regard to this circle, that is, these points are conjugate to one another in regard to this circle, and lie on a line passing through the centre of this circle. If we invert this system in regard to a circle, we obtain another circle and two points. The new points are inverse to one another in regard to the new circle.

For, every circle through the two original inverse points cuts at right angles the circle in regard to which they are inverse, as we have proved. Wherefore, by Ex. 2 above, every circle through the two new points cuts the new circle at right angles. This shews, by what we have said, that the two new points are inverse points in regard to the new circle.

Ex. 4. It follows, from Ex. 3, that, when a circle, say S, of centre C, inverts into a circle, S', the point C' which is the inverse of C (in regard to the circle of inversion), is the in- verse of the centre, O, of the circle of inversion, taken in

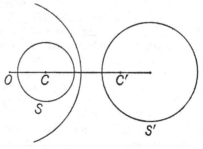

regard to S'. For, if K be any point of the line IJ, the points C

and K are inverses of one another in regard to S. The inverse of K (in regard to the circle of inversion) is, however, O. Thus O, C' are inverse points in regard to S'. In particular, when S' is a line, it passes through the middle point of OC' and is at right angles to this.

Examples in regard to circles. *Ex.* 1. If any three circles be taken, and the common chord, other than IJ, of each of the three pairs of these circles, these three common chords meet in a point. Let the circles be S, U, V; consider the common points of S with U, and, also, of S with V; let the former be P, P', and the latter be Q, Q', and let the chords PP', QQ' meet in T. By what has been seen, a circle can be described with T as centre, to cut the circle S at right angles; it is the conic touching TI and TJ, respectively at I and J, which passes through one (and therefore the other) of the points of contact of the tangents drawn from T to S. The points P, P', being on S, are inverse in regard to this circle, as also are Q, Q'. This circle, therefore, cuts U and V, also, at right angles. Its centre T is, therefore, on the common chord of U and V. For, if H be one of the common points of U and V, and TH meet U and V again, respectively, in L and M, it follows that L and M are both inverses of H, in regard to the circle of centre T, and, therefore, coincide.

Ex. 2. Let A, B, C be any three points; and D, E, F be any three points lying respectively on BC, on CA and on AB. Through

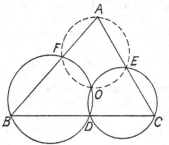

the three points F, B, D (beside I, J) a circle can be drawn; through the three points E, C, D another circle. These, beside D (and I, J), will have common a further point O. This point lies on the circle passing through A, E, F. (Miquel, *Liouv. J.*, iii, 1838.)

Denote the points in which the line IJ is met by BC, CA, AB, respectively, by A', B', C'; and the points in which the lines OD, OE, OF, respectively, meet IJ, by D', E', F'. Then the pairs of points I, J; A', F'; C', D' are in involution, that determined on IJ by the conics through the four points F, B, D, O; so that the range C', F', J, I is related to the range D', A', I, J. Again, by conics through E, C, D, O, the pairs I, J; A', E'; B', D' are in involution, and, therefore, the range D', A', I, J is related to the range B', E', J, I, and is, hence, related to the range E', B', I, J. But, then, the ranges, C', F', J, I and E', B', I, J, being related, the pairs I, J; B', F'; C', E' are in involution. This shews that a circle passes through the points A, E, F, O.

Ex. 3. Wherefore, by supposing the points *D, E, F* to be in line, we infer that, if four lines be taken in a plane, and, from every three of these lines, by inter-sections of pairs, a triad of points be defined, then the three circles deter-mined by three of these triads have a common point. Therefore, by sym-metry of reasoning, the four circles determined by the four triads have a point in common.

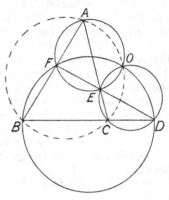

Conversely, let points, *D, E, F*, lying, respectively, on the three joins *BC, CA, AB*, be such that the circle *ABC* contains the point of concur-rence, *O*, of the three circles *AEF*, *BFD, CDE*, whose existence is proved in Ex. 2. Then the points *D, E, F* are in line. For, let the line *EF* meet *BC* in *P*. The point *O* is determined by the intersection of the circles *AEF* and *ABC*, and the circle *EPC* passes through *O*. But this circle has, with the circle *CDE*, the points *E, O, C, I, J* in common, and thus coincides with it. Hence *P* coincides with *D*.

The circle *ABC* may thus be regarded as the locus of the point of intersection of the circles *AEF, BFD, CDE*, when *D, E, F* are determined by a varying line.

Ex. 4. A particular corollary from this is the fact (already proved, Chap. I, Ex. 8) that, if, from a point, *O*, of the circle *ABC*, lines be drawn at right angles to *BC*,

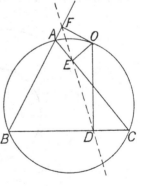

CA, AB, meeting these, respectively, in *D, E* and *F*, then these points *D, E, F* are in line. For, the pencil of lines *E* (*I, J, O, A*), being harmonic, is related to the pencil *F* (*I, J, O, A*). Thus a circle can be drawn through *O, E, F, A, I, J*. Similarly, circles can be drawn through *O, F, B, D* and *O, D, C, E*.

If the lines *OD, OE, OF* meet the line *IJ*, respectively, in *D', E', F'*; and *L* be the harmonic conjugate, on the line *OD*, of the point *O* in regard to *D* and *D'*, so that *D* is the middle point of *OL*; and, similarly, *M, N* be taken, on *OE, OF*, respectively, such that *E* is the middle point of *OM*, and *F* the middle point of *ON*, it follows that *L, M, N* are in line.

Ex. 5. If, in the preceding example, there be four lines, instead

72   *Chapter II*

of only the three *BC, CA, AB*, and *O* be the point of concurrence of
the four circles, determined, as in Ex. 3, by every three of these,
there will be obtained four points, *L, M, N, P*, lying in line. The
rule for these is, that the middle point of *OL* lies on one of the
four given lines, the line *OL* being at right angles to this line.

Now apply the theory of inversion, explained above, to this figure,
the circle of inversion being any circle of which *O* is the centre.
The four given lines then invert into circles passing through *O*;
the centres of these circles will (Ex. 4, above, in regard to inversion)
be the points obtained by inversion of *L, M, N, P*. These centres
will then lie on a circle passing through *O*. In the original figure,
the point *O* was obtained as a point of concurrence of four circles;
of these, every two have an intersection beside *O*, which is also an
intersection of two of the given lines. These circles will invert into
four lines, say *a', b', c', d'*. The original figure can, then, be obtained,
by inversion, from four arbitrary lines *a', b', c', d'*, and an arbitrary
point, *O*.

We thus infer a further property of the four circles considered in
Ex. 3 above (given, with other properties, by Steiner, *Ges. Werke*,
I, p. 223)—namely, that their centres lie on a circle which passes
through *O*. This result may, also, be regarded as built up from
the theorem, similarly obtainable from Ex. 4, by inversion, that,
if three circles with a common point be such that the other three
points of intersection of the pairs of these lie in line, then the
centres of the three circles lie on a circle passing through their
point of concurrence.

Ex. 6. In the figure of Ex. 3, we have four lines, and four circles
which meet in *O*. This arose from the figure of Ex. 2, in which there

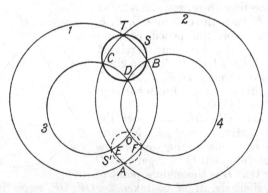

were three lines, and three concurrent circles. There is a generalisation, of importance, in which the lines are replaced by circles.

Let any four circles, which we suppose considered in a definite order, be denoted, respectively, by 1, 2, 3, 4. Let the pairs of intersections (other than *I*, *J*), of the successive pairs of these circles, be denoted, respectively, as follows:

for 1, 2 by *T*, *A*; for 2, 3 by *C*, *E*; for 3, 4 by *D*, *O*; for 4, 1 by *B*, *F*.

Then it can be shewn that, if the four points *T*, *C*, *D*, *B* lie on a circle, say *S*, the four remaining points, *A*, *E*, *O*, *F*, also lie on a circle, say *S′*.

This result is obtainable by inversion in regard to any circle of centre *T*, whereby we are led back to Miquel's theorem of Ex. 2. But, as will be seen below, in dealing with the relations of geometry in two and three dimensions, this is not the simplest way of regarding the theorem.

Ex. 7. Three circles, 1, 2, 3, meet in a point *O*. The tangents at *O*, respectively of the circles 2 and 3, meet the circle 1 in *Q* and *R*; the points of intersection of the circles are *A* or (2, 3), *B* or (3, 1), *C* or (1, 2). Prove that *OA*, *QB*, *RC* meet in a point. (Chap. I, Ex. 34.)

Ex. 8. If a pair of common chords of two conics, which meet in *E*, meet two common tangents of the conics, respectively, in *X* and *M*, the line *XM* passes through one of the two points, *F*, which, with *E*

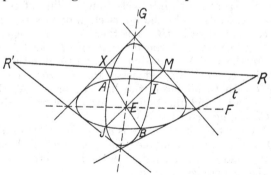

and another point *G*, form the common self-polar triad of the two conics. The points *X*, *M* belong to the involution determined on the line *XM* by the two conics, the double points of this involution being the point *F*, and the point of *XM* which lies on *EG*; the points, *R*, *R′*, in which the line *XM* meets the other two tangents common to the conics, also belong to this involution.

The result follows at once from the fact remarked (Chap. I, Ex. 14), that the figure consists of pairs of points which are harmonic images of one another in regard to the line *EG* and the point *F*.

Ex. 9. In the case of two circles, the common self-polar triad consists of the two limiting points, *F*, *G*, and the point, *E*, in which the common chord, *AB*, meets the line *IJ*. Let one of the four common tangents of the circles be called *t*, and consider, with this,

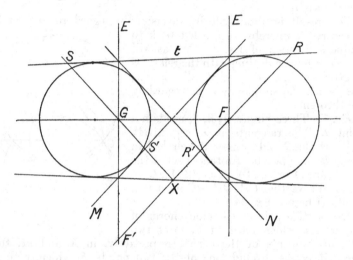

the common tangent which intersects *t* on the line of centres, *FG*, of the two circles; let *X* be the intersection of this latter tangent with the common chord *AB*. Another common tangent passes through the intersection of the line *EF* with *t*; let this meet the line *IJ* in *M*, the line *XM* passing through the limiting point *F*. The remaining common tangent passes through the intersection of the line *EG* with *t*, and meets the line *IJ*, say, in *N*. Let *XM* meet the tangent *t*, and the tangent through *N*, respectively in *R* and *R'*. (Cf. the figure of Ex. 8.)

If then *FX* meet *EG* in *F'*, there is on the line *FX* an involution of which *F* and *F'* are the double points, of which *X*, *M* and *R*, *R'* are two pairs, and two other pairs are the intersections of the line *XM* with the two circles. (Ex. 8.)

Now a circle, Ω, can be drawn with centre *X*, to contain the limiting points *F*, *G*, by what we have shewn above; this will cut both the given circles at right angles, and will, therefore, pass through the points of contact with the circles of the common tangent containing *X*. As *M* is on the polar, *IJ*, of *X*, in regard to this circle, and *X*, *M* are harmonic in regard to *F* and *F'*, this circle passes also through F'. Thus the points R, R' are inverses of one another in regard to this circle, Ω. In the same way, if XG meet the tangent t, and the tangent through M, respectively in S and S', it follows that S, S' are inverses of one another in regard to this same circle, Ω. The lines XF, XG, likewise, meet the given circles in points which are inverses of one another in regard to this circle, Ω.

If we now invert the figure in regard to this circle, Ω, both the given circles invert into themselves, but the common tangent t inverts into a circle, passing through the points X, R', S', which touches both the given circles.

Conversely, suppose that the three lines, which appear here as the common tangents containing respectively X, M, N, are given. It can be shewn that there are four circles touching these three lines; the proof has in fact been given (Chap. I, Ex. 17). From what is here shewn, it follows that these four circles are all touched by another circle. It can be shewn, moreover, easily, that the points X, R', S', through which this circle passes, are the middle points of the pairs of points formed by the intersections of the three given lines. This theorem has been proved above in another way (Chap. I, Ex. 33).

The present proof (usually based on metrical considerations here not utilised) was given by J. P. Taylor, *Quarterly Journal of Mathematics*, XIII (1875), p. 197; and by Fontené, *Nouvelles Annales*, VII (1907), p. 161. The theorem is considered again, in Chap. III, below.

Angle properties for a circle. In the present volume, no use is made of any measure of an angle until Chap. V is reached. In Euclid's treatment of circles great use is made of the fact that the angles in any segment of a circle are equal to one another; it will be convenient to give here a condition under which we may, if we desire, speak of two angles as being equal. In Euclid's theory, if two angles α, β be respectively equal to two angles α', β', then an angle, formed by the *sum* of the two former, is regarded as being equal to the angle formed by the *sum* of the two latter. This is a step which we do not take before Chap. V is reached, the *sum* of two angles being at present undefined.

Let OA, OB be two lines, meeting in a point, O; and let $O'A'$, $O'B'$ be two other lines, meeting in O'. We may say that, in regard to the fundamental points I, J, the line OB makes with OA an angle equal to that made with $O'A'$ by $O'B'$, if the pencil $O(B, A, I, J)$ be related to the pencil $O'(B', A', I, J)$. If OB, OA meet the line IJ, respectively, in Q and P, while $O'B'$, $O'A'$ meet IJ in Q' and P', respectively, the ranges (Q, P, I, J), (Q', P', I, J) are then related.

It may be remarked that, when we are dealing with real points,

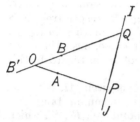

if B' be a point of the line OQ separated from B by O and Q, this provisional definition would render the angle made by OB' with OA equal to the angle made with OA by OB. These, however, are, in general, different from the angle made by OA with OB, or with OB'. The angle made by OB with OA is equal to the angle made by OA with OB, when OA, OB are harmonic conjugates in regard to I, J; that is, in the phraseology here used, when OA, OB are at right angles, in regard to I, J.

From this definition, if the points O, O', A, B lie on a circle, the angle which $O'B$ makes with $O'A$ is equal to the angle which

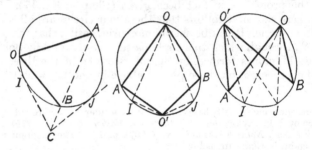

OB makes with OA. And, when the line AB passes through the centre, C, of the circle, in which case the lines joining O to A and B are harmonic conjugates in regard to OI, OJ, the lines OB, OA are at right angles.

Ex. If the tangents at two points, P, Q, of a circle meet in T,

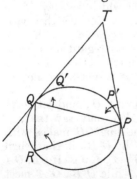

the angle made by QT with QP is equal to the angle made by PQ with PT; and, if R be any other point of the circle, this is equal to the angle made by RQ with RP.

For, if P', Q' be any other two points of the circle, the range P', Q', I, J, of four points of the circle, gives related pencils when joined to any two points of the circle. Thus, the pencils $Q(Q', P', I, J)$, $P(Q', P', I, J)$ are related. Recalling, what has appeared in the definition of a conic, that the tangent contains two coincident points of the conic, we see, by supposing P' to be at P, and Q' to be at Q, that the pencils $Q(T, P, I, J)$, $P(Q, T, I, J)$ are related. Further, the pencils

$Q(Q', P', I, J)$, $R(Q', P', I, J)$ are related, so that each of the preceding pencils is related to $R(Q, P, I, J)$.

Foci and axes of a conic. The *foci* of a conic, in regard to I and J, are defined as the two pairs of intersections, S, S' and H, H', of the two pairs of tangents to the conic from I, J. By the properties of conics touching four lines (above, p. 24), the three lines SS', HH', IJ form a self-polar triad, the intersection, C, of SS' and HH', being the pole of the line IJ in regard to the conic, that is, the *centre*, in regard to I and J. The lines CSS', CHH' meet IJ in two points which are harmonic conjugates in regard to I and J; these lines are called the *axes* of the conic; they are at right angles. The tangents CA, CB, to the conic,

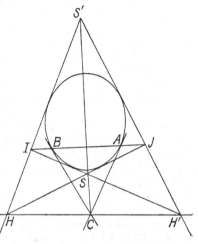

from the centre C, are called the *asymptotes*. Any two lines, drawn through the centre C, conjugate to one another in regard to the conic, that is, harmonic conjugates in regard to the asymptotes CA, CB, are called *conjugate diameters* of the conic. The axes are that single pair of conjugate diameters which are at right angles.

Ex. As a conic can be drawn to have five given tangents, a conic can be drawn to touch the joins BC, CA, AB, of three points A, B, C, and have a given point S for a focus. The tangents, other than IS, JS, drawn to this conic from I and J will intersect one another in a further focus, S', which we may call the focus conjugate to S.

We consider the correspondence of S and S', either of which determines the other. In particular we shew that if S be at the orthocentre of A, B, C, in regard to I, J, then S' is at the centre of the circle A, B, C.

The line pairs AI, AJ; AS, AS'; AB, AC are pairs of tangents drawn from A to particular conics touching the four lines IS, JS, IS', JS'; they are therefore in involution (Chap. I, Ex. 1). Similarly for the pairs of lines joining B to I, J; S, S'; C, A; and for the pairs of lines joining C to I, J; S, S'; A, B. Let the double rays of the first involution, of lines through A, intersect the double rays of the second involution, through B, in the four points X, Y, Z, T; then it follows, because the lines AYX, AZT,

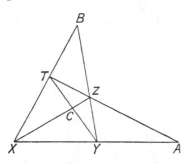

say, are harmonic in regard to AC, AB, and the lines BZY, BTX, say, are harmonic in regard to BA, BC, that the lines XZ, TY, say, meet in C. And, then, as the line IJ is divided harmonically by the line pair AYX, AZT, and also by the line pair BZY, BTX, it is divided harmonically by the line pair XCZ, YCT (above, p. 4); so, likewise, is the line SS'. Wherefore the lines XCZ, YCT are the double lines of the involution containing the line pairs CI, CJ; CS, CS'; CA, CB. Thus the relation of S, S' is that of two points which are conjugate to one another in regard to all conics passing through X, Y, Z, T (as in Ex. 16 of Chap. i), and I, J are conjugate to one another, also, in regard to such conics. Thus when S describes a line, S' describes a conic passing through A, B, C; and conversely.

A point D can be taken so that IJ is divided harmonically by AC, BD and by BC, AD; it will then (above, p. 4) also be divided harmonically by AB, CD. The point D will be the ortho-

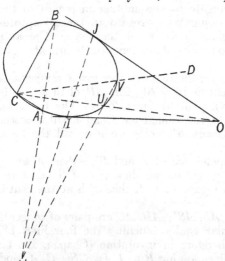

centre of A, B, C, in regard to I, J, and I, J will be conjugate to one another in regard to all conics through A,B,C,D. Let the conic (circle) be drawn through A, B, C, I, J, and let O be the intersection of the tangents of this conic at I, J, the centre of this conic. We shew then that when S is at D, then S' is at O. For let CO, CD meet the conic A, B, C, I, J, again, respectively in U, V, and let BA, JI meet in W. As the points I, J are separated harmonically by the lines AB and CD, while, as points on the conic, U and C are harmonic conjugates in regard to I and J, it follows that VU passes through W. Thus the line pairs *CI, CJ*; *CA, CB*; *CO, CD* are in involution. By a similar argument the line pairs joining *A* to the point pairs *I, J*; *B, C*; *O, D* are in involution; as, also, those joining *B* to *I, J*; *C, A*; *O, D*. This shews that *O, D* are related as was stated.

Confocal Conics. A system of conics having the same foci is called a confocal system. It is the same as a system of conics with four common tangents, of which two meet in the point *I*, and the other two in the point *J*. The four foci are in pairs, on two lines which meet in the centre of the conic; if one of these pairs be common to two conics, so are the other pair.

If the two pairs of foci be denoted, as above, by *S, S'* and *H, H'*; and if *PT, PT'* be the tangents, to one conic of the confocal system, drawn from an arbitrary point *P*, it is known (above, p. 25) that the four pairs of lines *PT, PT'*; *PI, PJ*; *PS, PS'*; *PH, PH'*, are in involution. The double lines of this involution are harmonic in regard to the lines of each pair; in particular in regard to *PI, PJ*, and so are at right angles. These double lines are, however, the tangents at *P*, of the two conics which can be drawn, through *P*, to touch the four lines *IS, IS', JS, JS'* (above, p. 25). Thus, through any point *P* can be drawn two conics of the confocal system, and these cut one another at right angles, at the point *P*. It can be shewn, similarly, that they also cut at right angles at each of their other three points of intersection.

Further, if we refer again to the provisional phraseology in regard to equal angles, the angle made by *PS'* with *PT'* is equal to the angle which *PT* makes with *PS*. For, as the pairs of lines, *PS, PS'*; *PT, PT'*; *PI, PJ* are in involution, the pencils *P*(*S', T', I, J*), *P*(*S, T, J, I*) are related, and the second is related to *P*(*T,S,I,J*). Also, if one of the double rays of this involution be *PM*, the pencils *P*(*M,S',I,J*), *P*(*S, M, I, J*) are related; so that the angle made by *PM* with *PS'* is equal to the angle made with *PM* by *PS*. In particular, when *P* is on the conic, the angles *S'PM, MPS* are equal.

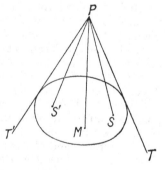

A particular set of confocal conics is a set of confocal parabolas, that is, conics all touching the line *IJ* at the same point, and having two other common tangents, *IS, JS*.

Common chords of a circle with a conic. If the axes of a conic meet the line *IJ* in *X* and *Y*, as the axes meet in the pole of the line *IJ* and are at right angles and conjugate in regard to the conic, the points *X, Y* are the double points of the involution determined by the two pairs consisting of I, J and the points, F, F', in which the conic meets I, J ; this is the involution determined on the line IJ by the conic and any circle. If, then, a pair of chords, LM, $L'M'$, of the conic and the circle, meet IJ in P and P', the pair P, P' will belong to this involution. Wherefore, the ranges, P, X, I, J and X, P', I, J, are related. Thus, the angle made by the common chord, LM, of a conic, and any circle, with either axis, CX, is equal to the angle made by this axis with the complementary common chord, $L'M'$, of the conic and circle.

The director circle of a conic. The director circles of conics touching four lines are coaxial. This is the conic which has been defined (Chap. I, Ex. 3) as the director conic in regard to I and J. From its definition it is the locus of points from which the two tangents to the conic are at right angles ; it passes through the points in which the conic is met by the polar of any one of the four foci; and its centre coincides with the centre of the conic. As has been shewn (Chap. I, Ex. 25), if, through three points which form a self-polar triad in regard to the conic, the circle be drawn, this cuts the director circle at right angles.

Further, the director circles of all conics which touch four given lines have two points in common, beside I and J, that is, are a coaxial system of circles. For, let P be one of the intersections of the director circles of two particular conics touching these four lines. The pairs of tangents from P, to all these conics, are a pencil in involution (above, p. 25); and these meet the line IJ, which is supposed to have no special relation to the four lines, in pairs of points in involution. Of this involution, however, the points I, J are the double points; for, the tangents from P to each of the two selected conics are at right angles, and an involution is determined by two of its pairs. Wherefore, the tangents from P to any other of the conics are at right angles, and P lies on the director circle of this conic. From this, the statement made follows.

Particular, degenerate, conics touching the four lines, are the point pairs, three in number, through which the lines pass in pairs. If A, C be such a pair, the lines PA, PC are the tangents to this from a point P, and the corresponding director circle is the locus of P when the lines PA, PC are at right angles. The centre of this circle is the middle point of AC in regard to I and J. If B, D, and E, F, be the other point pairs, the middle points of AC, BD, EF lie on a line, the line containing the centres of all the coaxial director circles, which are the centres, the poles of IJ, in regard to the conics touching the four lines.

One conic can be drawn touching the four lines which also touches IJ. The director circle of this, as was remarked, degenerates into the line IJ and another line; if the other two tangents which can be drawn, from I, J, to this conic, meet in S, the line in question is the polar of S in regard to this conic. (Cf. Ex. 5 of Chap. I.) This line is then the radical axis of all the director circles, and is at right angles to the line of centres of the conics.

Parabolas. The directrix of a parabola. A conic touching the line IJ is called a *parabola*. If this touch IJ in S', and if S be the intersection of the tangents from I, J, the four foci of the parabola are S, S' and I, J. The polar of S in regard to the parabola is called the *directrix*; it was this name which suggested, to Gaskin, to call the director circle of any conic, the *director*. As a particular case of what has been said, since a line cuts a circle at right angles only if it passes through the centre of the circle, it follows that the circle drawn through three points, which are a self-polar triad in regard to a parabola, has its centre on the directrix of the parabola.

The directrix of a parabola also passes through the orthocentre of any three points whose joins are three tangents of the parabola. This follows at once from Brianchon's theorem. For let A, B, C be three points such that BC, CA, AB touch the parabola: let CB, CA, respectively, meet the directrix in P and Q; let the other tangents from P, Q, to the parabola, meet IJ, respectively, in T and U. As then, by definition, the tangents, PT, CB, are at right angles, the line AT is the line drawn from A at right angles to BC. So BU is the line drawn from B at right angles to AC. The point of intersection of these lines is the orthocentre of A, B, C (above, p. 64); by Brianchon's theorem (Chap. I, Ex. 2) this point of intersection is on

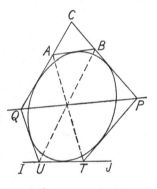

the directrix PQ. Conversely, it may be seen that a parabola can be drawn, to touch three given lines, to have for directrix any line through the orthocentre of the points formed by the intersections of the lines.

Examples in regard to a parabola. *Ex.* 1. If A, B, C be three points such that the lines BC, CA, AB touch the parabola, the circle through A, B, C contains also the focus, S, of the parabola. For, as the lines joining S to the points I, J are, by the definition of S, tangents of the parabola, and the line IJ is also a tangent, this is the theorem (Chap. I, Ex. 7) that, when the six lines, which are the joins of the pairs of each of two triads of points, touch a conic, the two triads are points of another conic.

Ex. 2. If we have four lines, any three of these, by their intersections in pairs, give rise to a triad of points, through which there passes a circle. The four circles so obtained meet in a point, namely the focus of the parabola which can be drawn to touch the four lines (beside the line *IJ*). This is a result obtained above (p. 71). It will be seen later that the theorem is equivalent to Moebius' theorem of the in- and circumscribed tetrads (Vol. I, p. 62).

A conic, *S*, containing a triad of points whose joins touch another conic, Σ (and, therefore, an infinite number of such triads), may be spoken of as triangularly circumscribed to Σ. Thus, any circle passing through the focus of a parabola is triangularly circumscribed to the parabola.

Ex. 3. Let the line joining the focus, *S*, of a parabola, to the point, *S′*, in which the curve touches *IJ*, meet the curve again in *A*. Then *A* may be called the vertex. It can be shewn that the line drawn through the focus, *S*, at right angles to any tangent of the curve, meets this tangent at a point lying on the tangent at the vertex, *A*.

For, take any point *P* of the directrix, the polar of *S*. Draw from *P* the two tangents of the curve; let these meet the line *IJ*,

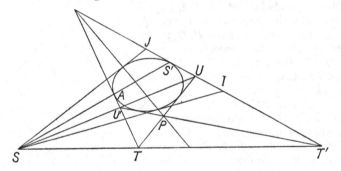

respectively, in *U* and *T′*, and meet the tangent at *A*, respectively, in *T* and *U′*, so that, as is easy to prove, *S*, *U*, *U′* are in line, and *S*, *T*, *T′* are in line. Then *U*, *T′* are harmonic conjugates in regard to *I* and *J*. That is, the line *ST*, drawn from *S* at right angles to the tangent *TU*, meets this tangent in a point, *T*, of the tangent at the vertex.

Combining with the last example, we infer that the lines drawn from the Miquel point of four lines, there obtained, each at right angles to one of the lines, meet these lines respectively in points which are in line. Also, the orthocentres of the four triads, determined by each set of three of the lines, are in line. Further, the pedal line of a point, *O*, of a circle, in regard to three points of the circle, of which the orthocentre is *K*, passes through the middle point of *OK*. (Cf. Ex. 8, of Chap. I; and Ex. 4, above, regarding the circle.)

Ex. 4. Any line drawn through the orthocentre, *K*, of three points, *A*, *B*, *C*, of a circle, meets *BC*, *CA*, *AB*, respectively, in the points *D*, *E*, *F*, and the lines *AK*, *BK*, *CK* meet the circle again, respectively, in *P*, *Q*, *R*. Prove that the lines *DP*, *EQ*, *FR* meet in a point, *S*, of the circle. And that this is the focus of the

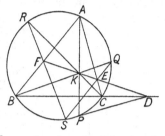

parabola which, as has been remarked, can be described to touch *BC*, *CA*, *AB* and have the given line as directrix.

Ex. 5. The joins of three points, *A*, *B*, *C*, which form a self-polar triad in regard to a conic, meet any tangent of the conic, respectively, in *D'*, *E'*, *F'*; on these joins are taken *D*, *E*, *F* such that *D*, *D'* are harmonic conjugates in regard to *B*, *C*, and so on. Prove that the lines *EF*, *FD*, *DE* are tangents of the conic: Thus, if *A, B, C* be a self-conjugate triad in regard to a parabola, the circle

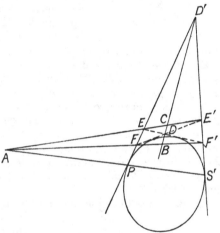

through the middle points of *BC*, *CA*, *AB*, contains the focus of the parabola.

Definition of a rectangular hyperbola. A conic which meets the line *IJ* in two points which are harmonic conjugates of one another in regard to *I* and *J*, is called a *rectangular hyperbola* in regard to *I* and *J*. In particular, a degenerate rectangular hyperbola consists of two lines which are at right angles. A conic which touches the line *IJ*, at one of the two points *I, J*, is both a parabola and a rectangular hyperbola.

By considering the involution of pairs of points determined on the line *IJ* by the conics which pass through the four intersections of two rectangular hyperbolas, we see that all these conics are rectangular hyperbolas. In particular, if *K* be the orthocentre of a triad of points, *A*, *B*, *C*, all conics through *A*, *B*, *C*, *K* are rectangular hyperbolas; conversely, if any rectangular hyperbola drawn through A, B, C meet the line, which is drawn through A at right angles to BC, in the further point K, the line BK is at right angles to CA; thus any rectangular hyperbola drawn through A, B, C also contains the orthocentre.

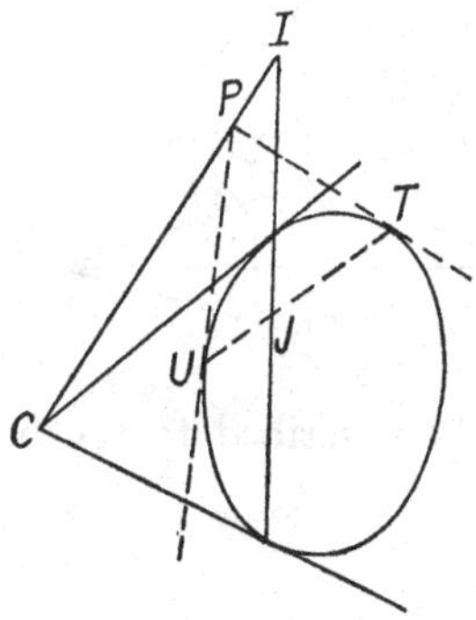

The centre, C, of a rectangular hyperbola, and the points I, J, form, by definition, a self-polar triad in regard to the rectangular hyperbola. The lines PI, PJ, when P is any point of the line CI, are thus conjugate in regard to the curve, and are, therefore, harmonic in regard to the tangents, PT, PU, which can be drawn from P to the curve. These tangents are therefore at right angles. Thus we see that the director circle of a rectangular hyperbola degenerates into the line pair joining its centre to the points I and J.

Examples in regard to a rectangular hyperbola. *Ex.* 1. Any pair of lines CP, CD, drawn through the centre, C, of a rectangular hyperbola, which are conjugate lines in regard to the curve, are harmonic in regard to the tangents of the curve, CL and CM, at the points, L and M, where the curve meets the line IJ. In particular CI, CJ are such a pair of conjugate lines. If CP meet the curve in P, and the tangent at P meet CL, CM in L' and M', then P is the middle point of $L'M'$ in regard to IJ. Also the angle which CD makes with CL is equal to the angle which CL makes with CP.

Ex. 2. If the lines joining any point, P, of a rectangular hyperbola, to the points I and J, meet the curve again, respectively, in A and A', the line AA' passes through the centre, C, of the curve, and is at right angles to the line CP. In fact the lines $A'I$, AJ, PC meet in a point of the curve.

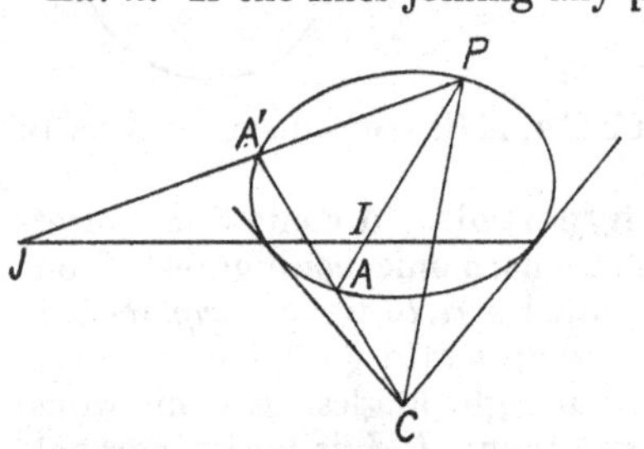

Further, if the tangents of the conic at A and A' intersect in T (which will lie on IJ), and any line through T meet the rectangular hyperbola in M and N, prove that PM, PN are at right angles, in regard to I and J.

In general, for any conic, not necessarily a rectangular hyperbola, if through a fixed point, P, of the conic, lines be drawn at right angles, meeting the conic again, respectively, in M and N, the pairs M, N, so arising, belong to an involution of points of the conic.

Thus MN passes through a fixed point. It can be seen that this is on the line, drawn through P, at right angles to the tangent at P.

Ex. 3. The centre of a given rectangular hyperbola is R, and O is any point of the curve. A circle is drawn, with centre O, to cut the curve in A, B, C, D. If we form the polar reciprocal of the rectangular hyperbola in regard to the circle, that is, the envelope of the polars, in regard to the circle, of the points of the rectangular hyperbola, prove that the curve obtained is a parabola touching the tangents of the circle at A, B, C, D, whose focus is the pole, in regard to the circle, of the line through R at right angles to OR, that is, the inverse of R in regard to the circle, the directrix of the parabola passing through O.

Auxiliary circle of a conic. We now explain a method by which we may pass from any conic to another, or conversely from this other to the original, the deduction in the latter case being exactly the dual of the former.

Let Ω be any conic, and C, S any two fixed points, the line CS meeting Ω in A and A'. Let R be any variable point of Ω. Let CR and SR, respectively, meet Ω again in Z and Z'.

As then Z, R are pairs of an involution of points on the conic Ω, the range (Z) on the conic is related to the range (R). This in turn is, similarly, related to the range (Z'). Wherefore, by a result proved above (Ex. 28, Chap. I), the line ZZ' touches another conic, say, Σ. The points A, A' are the common corresponding points of the two ranges (Z), (Z'), and the conic Σ touches Ω at A and A'. If P be the point of contact of ZZ' with Σ, and T the intersection of ZZ' with AA', it is easy to see that P, T are harmonic, both with respect to Z, Z', and also with respect to the points in which ZZ' meets the tangents at A and A'.

Now denote by c the polar of C in regard to Ω, and by s the polar of S; let c meet Ω in I, J, and s meet Ω in H, K. It can at once be proved that the conic Σ touches CH, CK at the points where these meet the line c, and touches SI, SJ at the points where these meet the line s.

If the line ZZ', which we denote by r, meet the lines c, s, respectively, in U and V, and RU, RV meet the conic Ω in L and M, respectively, C, L, Z' will be in line, as, also, S, M, Z will be in line. The line RL is then a tangent of the conic Σ, being the position of ZZ' when R is at Z'; for a like reason, RM is a tangent of Σ.

Hence, if the conic Σ be given, and the lines c and s, the conic Ω can be defined by finding the points, U, V, where a variable tangent, r, of Σ, meets c and s, drawing from U, V the tangents (other than r) UL, VM, and finding the locus of their intersection, R. This is exactly the dual of the construction by which Σ is found when Ω

and the points C, S, are given. Further, as the line $Z'L$ passes through the pole of IJ in regard to Ω, the points Z', L are, as points of the conic Ω, harmonic conjugates in regard to the points I, J of Ω; thus, the line SR is the line through S drawn to the point, of the line IJ, which is the harmonic conjugate, in regard to I and J, of the point, U, where the tangent RL, of Σ, meets

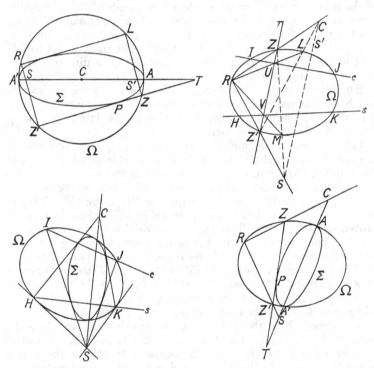

the line IJ. If we suppose the points S, I, J to be given, this shews how we may determine either of the conics Σ, Ω when the other is given. The lines $Z'R, LZ$ meet on IJ; thus, if we take the point, S', of SC, which is the harmonic conjugate of S, in regard to C and the point where SC meets IJ, then the lines $S'LZ$ and RL will meet the line IJ in points which are harmonic conjugates of one another in regard to I and J.

If we now regard I, J as the absolute points, the conic Ω will be a circle, of which C is the centre. The conic Σ will then also have C for centre, and S, S' for foci. The circle, Ω, is, then, the locus of the foot of the perpendicular drawn from (S or from S'), to a tangent of Σ; or, conversely, Σ is thereby determined from Ω.

The circle Ω is that generally called the auxiliary circle of the conic Σ.

Ex. 1. If we be given a fixed point, S, and a line; and from a variable point, N, of this line, the line be drawn at right angles to SN, then this line envelopes a parabola, which touches the given line.

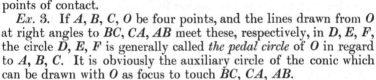

Ex. 2. If two conics, Σ, Ω, have two points of contact, and a variable tangent of one of these, Σ, meet Ω in P and Q, the further tangents from P and Q, drawn to Σ, meet in a point T, whose locus is a further conic touching both of Σ and Ω at their two points of contact.

Ex. 3. If A, B, C, O be four points, and the lines drawn from O at right angles to BC, CA, AB meet these, respectively, in D, E, F, the circle D, E, F is generally called *the pedal circle* of O in regard to A, B, C. It is obviously the auxiliary circle of the conic which can be drawn with O as focus to touch BC, CA, AB.

Let any conic be given, on which O is a fixed point, and A, B, C are variable points. We prove that the pedal circle passes through a fixed point.

For this, let the lines OI, OJ meet the given conic again, respectively, in M and N. There is then, as the six points O, M, N and A, B, C are on the conic, another conic touching BC, CA, AB, OM, ON and MN (Chap. i, Ex. 7). This will be the same as the conic having O for focus which touches BC, CA, AB; which conic, therefore, touches MN. Therefore, the pedal circle of O in regard to A, B, C passes through the fixed point in which MN is met by the line drawn from O at right angles to MN.

In particular, we have remarked (Ex. 2, p. 84) that, when the conic is a rectangular hyperbola, this point is the centre. Thus the pedal circle of any point of a rectangular hyperbola, in regard to any other three points of the curve, passes through the centre of the rectangular hyperbola.

When the conic is a circle, or passes through the points I, J, it has been seen (Chap. i, Ex. 8, and p. 71, above) that the pedal circle, defined with reference to I and J, degenerates into a line, together with the line IJ. It can, however, be shewn, in this case (see below, Chap. iii, Ex. 20), that, if we consider three variable points, A, B, C, of the circle, which are a self-polar triad in regard to another fixed conic (which must then be taken so that this is possible, being inpolar to the circle; see Chap. i, Ex. 11), then the pedal line of a fixed point of the circle in regard to these points A, B, C, passes through a fixed point.

The pedal circle of a point O, in regard to three points A, B, C, being the auxiliary circle of a conic, having O for focus, which touches *BC*, *CA*, *AB*, also arises as the pedal circle, in regard to *A*, *B*, *C*, of another point, *O'*, the conjugate focus of this conic. We have shewn above (p. 78), that when *O* describes any conic passing through *A*, *B*, *C*, then *O'* describes a line; and that, when *O* is at the orthocentre of *A*, *B*, *C*, then *O'* is at the centre of the circle through *A*, *B*, *C*. Thus, when the point *O* is on a rectangular hyperbola through *A*, *B*, *C*, which then also passes through the orthocentre (above, p. 84), the point *O'* lies on a line through the centre of the circle *A*, *B*, *C*. We thus have the result, that, if a point *O'* describe any definite line passing through the centre of the circle *A*, *B*, *C*, the pedal circle of *O'* in regard to *A*, *B*, *C*, passes through a fixed point, which is in fact the centre of the rectangular hyperbola through *A*, *B*, *C* which corresponds to the line. It will be seen, below, that this point is the middle point, in regard to *I*, *J*, of the two points consisting of the orthocentre, and the fourth intersection of the hyperbola with the circle *A*, *B*, *C*.

Ex. 4. If a circle be described, with centre at a point, *O*, of a rectangular hyperbola, to meet this in *A*, *B*, *C*, *D*, and *DO* meet the circle again in *S*, prove that the pedal line of *S*, in regard to *A*, *B*, *C*, passes through the middle point of *O*, *S*. Hence shew that the directrix of the parabola, described with focus at *S* to touch *BC*, *CA*, *AB*, passes through *O*; and obtain this result by the transformation referred to above.

Ex. 5. Prove that to any point *P* of the conic Σ there corresponds a point *P'* of the conic Ω such that *CP'*, *SP* meet on the line *IJ*, while *CP*, *SP'* meet on the line *HK*. Taking two arbitrary points, *C*, *S*, and two arbitrary lines *IJ*, *HK*, investigate the correspondence between an arbitrary point *P* and the point *P'* determined by these two conditions. (Boscovich, *Sectionum Conicarum Elementa nova quadam methodo concinnata*, *Elementa Universae Matheseos*, Venetiis, 1757. See Taylor's *Ancient and Modern Geometry of Conics*, p. 3, etc.)

The locus of the centres of rectangular hyperbolas passing through three given points. We considered in Chap. i (Ex. 18) the locus of poles of a given line in regard to all conics passing through four given points, say, *A*, *B*, *C*, *D*. We found that it is a conic whose intersections with the line are two points, *I* and *J*, which are conjugate in regard to all conics passing through *A*, *B*, *C*, *D*; and we specified nine other points thereon, namely, first, the three intersections of the line pairs passing through these four points, and, second, the points, each on the join of two of these points, harmonic, in regard to these two points, to the point in which the join meets the given line *IJ*. When we take the points *I*, *J* as absolute points, each of the points *A*, *B*, *C*, *D* is the orthocentre of the other three, and the conics passing through these four points are rectangular hyperbolas. We thus have the result that, the locus of the centres of the rectangular hyberbolas which pass through three given points *A, B, C* (and therefore, also, through the orthocentre, *D*, of these) is a circle, which contains the three points where *AD, BD, CD*, respectively, meet *BC, CA, AB*, and contains, also, the middle points, in regard to the line *IJ*, of each of the six pairs of points, *BC, CA, AB, AD, BD, CD*. This is usually called the nine points circle of *A, B, C*. It is, however, equally the nine points circle of any three of the four points *A, B, C, D*.

The nine points circle can however be defined as the locus of the middle point, in regard to the line *IJ*, of the points *D, Q*, where *Q* is any point of the circle *A, B, C*, the point *D* being the orthocentre of *A, B, C*.

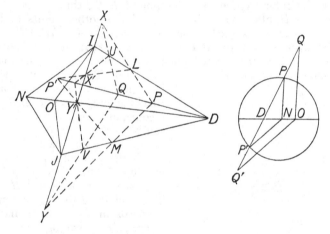

It is, in fact, the case that if a variable line be drawn, through a fixed point, *D*, to meet a given circle in *P* and *P'*, and a point *Q* be taken on *DP*, such that *P* is the middle point of *DQ* in regard to the line *IJ*, then the locus of *Q* is another circle. This contains the point, *Q'*, of the line *DPP'*, which is such that *P'* is the middle point of *DQ'*; and, if the centre of the given circle be *N*, the tangents of the new circle at *I, J* meet in a point *O* on *ND* such that *IO, ID* are harmonic in regard to *IJ* and *IN*, namely *N* is the middle point of *DO*. To prove the result, let *DI, DJ* meet the original circle in *L, M*, respectively; take *U* harmonically conjugate to *D*, in regard to *I* and *L*, and take *V* similarly on *DJ*, in regard to *J* and *M*. Then *QU, PL* meet, in *X*, on *IJ*; and *QV, PM* meet, in *Y*, on *IJ*. Thus the pencil, of fixed centre *U*, denoted by *U(Q)*, is related to the range *(X)*, and, thence, to the pencil *L(P)*; this last is related to $M(P)$, and, thence, to $V(Q)$. Thus the locus of Q is a conic containing U and V. In the same way UQ', LP' meet, in X', on IJ, and VQ', MP' in Y'; and the proof of the statement may be completed.

If the given circle be the nine points circle of A, B, C, of which D is the orthocentre, the point Q, as P describes the nine points circle, describes a circle. We have seen however that the nine points

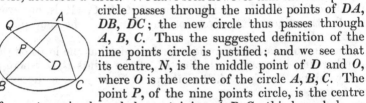

circle passes through the middle points of DA, DB, DC; the new circle thus passes through A, B, C. Thus the suggested definition of the nine points circle is justified; and we see that its centre, N, is the middle point of D and O, where O is the centre of the circle A, B, C. The point P, of the nine points circle, is the centre of a rectangular hyperbola containing A, B, C; this hyperbola, as it contains D, also passes through Q. Three other points of this hyperbola can be assigned, lying respectively on AP, BP, CP.

The polars of a point in regard to a system of confocal conics touch a parabola. The theorem obtained from the dually corresponding figure deserves mention also. The polars of a point

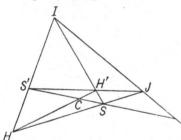

in regard to a system of conics touching four given lines envelope a conic, of which eleven tangents can be easily assigned.

Let the point pairs through which the four lines pass be the absolute points I, J, together with the pairs S, S' and H, H'. The conics are then a system of confocals in regard to I and J, whose pairs of conjugate foci are S, S' and H, H', their common centre being the intersection, C, of the axes SS' and HH'. Denote the fixed point by O.

Then three of the eleven tangents of the envelope are, by what has been seen above, the line IJ, and the lines CS, CH; thus the envelope is a parabola touching the axes of the confocal conics. And, as these axes are at right angles, in regard to I and J, the directrix of the parabola passes through the common centre, C. Two other tangents of the envelope are the tangents at O of the two confocals which pass through O; these tangents, being harmonic in regard to OI, OJ, are at right angles, and the directrix of the parabola passes through O. Other six tangents of the envelope are lines passing through the points I, J, S, S', H, H', in which the four given lines intersect; to find the line in question through one of these six points, we are to join this point to O, and take the harmonic conjugate of this join, in regard to the two of the four given lines which meet in the point selected. In particular, let *Q* be the point such that, *IQ, IO* are harmonic in regard to *IS, IH*, while *JQ, JO* are harmonic in regard to *JS', JH*; then *QI, QJ* are tangents of the envelope, and *Q* is the focus of this. This point *Q* is the conjugate of *O* in regard to all conics through the four foci *S, S', H, H'*.

Now let Σ denote a particular one of the confocal conics, and let one of the four common tangents, of Σ and the parabola, touch Σ in *P*; as this tangent touches the parabola there is one of the confocal conics in regard to which *O* is the pole of this line. On the other hand, the poles of this tangent in regard to all the confocals lie on a line, through *P*, at right angles to the tangent, this new line being conjugate to the tangent in regard to all the confocals, and, in particular, in regard to the point pair *I, J*. (Chap. i, Ex. 15.) Wherefore, *OP* is the line drawn through *P* at right angles to the tangent at *P*, which is called the normal at *P*. It is thus possible to draw through *O* four normals to the particular conic Σ.

If we take the poles, in regard to this particular conic Σ, of all the tangents of the parabola, these lie upon another conic, the so-called polar reciprocal of the parabola in regard to Σ. This new conic evidently contains the four points of Σ, such as *P*, whereat the normal of Σ passes through *O*. It is easy to see that it is a rectangular hyperbola, containing the centre *Ć* of all the confocal conics, and the point *O*, and the points of the line *IJ* which lie upon the axes, *CS, CH*, of the confocal conics. The lines *CS, CH, IJ* are in fact tangents of the parabola, and their poles, in regard to Σ, are three of the points spoken of; the point *O* is, by definition of the parabola, the pole of a tangent of this, in regard to Σ. That the four normals to a conic, Σ, from a point *O*, can be constructed by this rectangular hyperbola, was known to Apollonius (Conics, Lib. v, Props. 58–63); we consider the theorem again below (p. 93).

A particular theorem. For clearness we give here a particular result which will be found to be of interest.

Let *A, B, C, D* be any four points, and *X, Y* two points which are conjugate to one another in regard to all conics passing through these. Take one such conic, say, Σ, and let *T* be the pole of *XY* in regard to Σ. Let *TD* meet Σ again in *D'*. We prove that the six points *A, B, C, X, Y, D'* lie on a conic. The theorem has already been proved

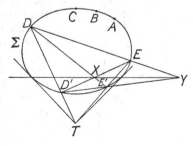

(above, p. 89). For, with X, Y as the absolute points, the point D is the orthocentre of A, B, C, and the conic Σ is a rectangular hyperbola of which T is the centre; the conic A, B, C, X, Y is the circle A, B, C, which was proved to meet the rectangular hyperbola in a point D', such that T is the middle point of DD' in regard to the line XY, as here. But we give another proof. If the lines YD, YD' meet the conic Σ, again, in E and E', respectively, the lines DD', EE' meet on the polar of Y, in T; and the lines DE', $D'E$ meet on the polars of Y and T, at X. As DB, AC meet the line XY in two points harmonic in regard to X and Y, as, likewise, do DC, AB, the pencils, $D\,(X, Y, B, C)$ and $A\,(X, Y, C, B)$, are related; the former is the pencil $D\,(E', E, B, C)$, which is related to the pencil $D'\,(E', E, B, C)$ or $D'\,(Y, X, B, C)$, which is related to $D'\,(X, Y, C, B)$. This is therefore related to $A\,(X, Y, C, B)$. Thus the points D', A, X, Y, C, B lie on a conic.

Conversely, let two conics, Σ, Σ', intersect in A, B, C, D', and a chord, XY, of one conic, Σ', meet the other conic, Σ, in two points harmonic in regard to X and Y; and let the line joining the pole, T, of this chord, in regard to Σ, to one of the four common points, D', of the two conics, meet Σ again in D. Then the points X, Y are conjugate to one another in regard to all conics through A, B, C, D. Wherefore, if I, J be any two points of the line XY which are harmonic in regard to X and Y, a conic can be drawn through A, B, C, D, I, J.

From this, if we regard I, J as the absolute points, in which case the conic Σ' is a rectangular hyperbola, we have the theorem that, if a conic, Σ, whose centre is T, be intersected by a rectangular hyperbola, Σ', which meets the line IJ in two points which are conjugate in regard to the conic Σ, in A, B, C, D'; and D be the other

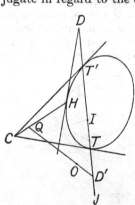

point in which TD' meets Σ; then the four points A, B, C, D lie on a circle. We shall see that this result is, particularly, of interest when A, B, C are such that the rectangular hyperbola contains the centre, T.

The normals of a conic which pass through a given point. Let O be any fixed point, and H a variable point of a given conic. Let C be the centre of the conic. If through O the line be drawn which is at right angles to the tangent at H, and this line meet CH in Q, we consider the locus of Q as H varies. If QO meet the line IJ in D', and the tangent at H meet the line IJ in D, so that D, D' are harmonic conjugates in regard to I and J, the pencil $O(Q)$, or $O(D')$, is related to the range (D), on the line IJ. The line CH, through the pole, C, of the line IJ, and through the point of contact, H, of a tangent from D, is, however, the polar of D; the line CH meets IJ in a point which is the harmonic of D in regard to the points, T and T', in which the line IJ meets the conic. Thus the pencil $C(Q)$, or $C(H)$, is related to the range (D), and, hence, to the pencil $O(Q)$. Therefore the locus of Q is a conic, passing through the centre C, and the point O. If the axes of the conic meet IJ in A and B, these points are harmonic conjugates both in regard to T, T' and in regard to I, J. Thus, when D is at A, the line CH goes through B, and the point D' is at B; in this case Q is at B. The conic locus of Q is thus a rectangular hyperbola, containing both A and B.

At a position of Q which is on the given conic, and coincides with H, the line OQ is at right angles to the tangent of the given conic at Q, namely is the normal at Q. The four normals of the given conic, which pass through O, are then the normals of this conic at the points where it is met by the rectangular hyperbola.

It follows from what is said in the last section that, if these be the points Q_1, Q_2, Q_3, Q_4, and CQ_4 meet the given conic again in Q', then the points Q_1, Q_2, Q_3, Q' lie on a circle.

The preceding construction remains valid when the given conic is a parabola, touching the line IJ at C. If then, S being the focus, the line SC meet the parabola again in A, and the tangent at A meet the line IJ in B, the locus of Q passes through O and C, but also passes through B, and has the line CS for its tangent at C. As B is the pole of CS, and C, B are harmonic conjugates in regard to I, J, the locus of Q is still a rectangular hyperbola. To see that it passes through B, we 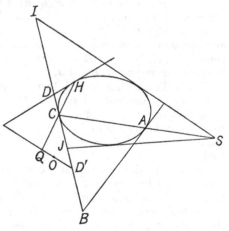 have only to take H at C; then D is at C, and D' is at B; to see that it has CS for its tangent at C, we have only to take H at A, in which case D is at B and D' is at C; so that the ray CA of the pencil $C(Q)$ corresponds to the ray OC of the pencil $O(Q)$. In this case, one of the four normals through O is always OC; and there are three others.

If OP, OQ, OR be these normals, it can be shewn, in this case, that the circle P, Q, R passes through A. For take P' on the given

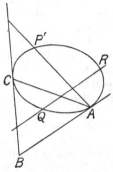

parabola, such that AP' and QR meet the line IJ in points which are harmonic in regard to C and B. As the parabola itself, touching IJ at C, has also this property, it follows, by considering the involution on the line CB by all conics through A, Q, R, P', that each of the line pairs AQ, RP' and AR, $P'Q$ also meets IJ in a pair of points which are harmonic in regard to C and B. Thus the pencil $P'(C, B, R, Q)$ is related to the pencil $A(C, B, Q, R)$; this, however, is related to the pencil of lines joining any point of the parabola to C, A, Q, R, and, thus, to the pencil $C(B, A, Q, R)$, and,

therefore, to the pencil $C(A, B, R, Q)$. As, then, the pencils $P'(C, B, R, Q)$, $C(A, B, R, Q)$ are related, we can infer that a conic constructed to pass through P', Q, R, C, B has CA for its tangent at C. Conversely, then, the rectangular hyperbola containing the points of the parabola whereat the normals pass through O, having been shewn to contain Q, R, C, B and to have CA for its tangent at C, must pass through P'; so that P' is the same as P. We thus infer that any conic through P, Q, R, A meets the line IJ in points which are harmonic in regard to C and B; or, conversely, that the conic through P, Q, R, and any two points of the line CB which are harmonic in regard to C and B, passes through A. Of such conics, the circle P, Q, R is one; for we have seen that C, B are harmonic in regard to I and J.

Ex. 1. Prove that an arbitrary given line is a normal of one conic of a given confocal system. If this line be normal at P, and meet this conic again in P', and if it meet another conic of the confocal system at Q and Q', prove that the normals of this other conic, at Q and Q', meet on the tangent at P', of the former conic.

Ex. 2. The dual conception to a normal of a conic is found by taking two arbitrary fixed lines i and j, and finding, upon a variable tangent of the conic, which touches this at P, the point P', which is the harmonic conjugate of P in regard to i and j. Prove that the locus of P' is a curve which meets an arbitrary line in four points, passes twice through the point of intersection of the lines i, j, and twice through each of two other points, which lie on the polar, in regard to the conic, of this point of intersection.

The dual of the director circle of a conic. The tangents to a conic from a point of the director circle are such as meet the fixed line IJ in a pair of an involution of points on this line, of which I, J

are the double points. Dually, if, from a given point, be drawn a pair of lines, belonging to an involution, and one of the points of intersection, with a given conic, of one of these lines, be joined to one of the intersections of the other line, then the join of the two intersections envelops another conic. If i, j be the double rays of the involution, the tangent of this second conic is any line whose intersections with the given conic are harmonic in regard to its intersections with the given lines i and j.

In fact both the constructions, that for the director circle, and that we have now described, arise in the same figure. Let the points P, Q, of a given conic, Ω, be such that the lines, OP, OQ, joining them to a fixed point O, are a pair of a pencil in involution, of which the lines, i, j, passing through O, are the double rays. Thus the line PQ is met by the lines i, j, respectively, in two points which are conjugate in regard to the conic Ω. Now, to any point on the line i, there may be made to correspond the point of the line j, which is

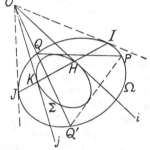

conjugate to the former in regard to the conic Ω; as these two points describe, on i and j, respectively, two ranges which are related, the line which joins them envelops a conic, say, Σ. This conic touches the lines i and j, say, in H and K, respectively. It also touches the four tangents, of the conic Ω, at the points where Ω is met by the lines i and j. If OQ meet Ω again in Q', the conic Σ also touches PQ'. And, if the other possible tangents be drawn to Σ from Q and Q', these meet, on the conic Ω, in the further point of the line OP which lies on Ω. Further, by considering the position of the line PQ in which it coincides with the line i, we see that O, H are conjugate to one another in regard to Ω; as, likewise, are O and K; thus HK is the polar of O in regard to Ω, meeting Ω at points, I, J, whereat the tangents pass through O. From this we infer that the lines PI, PJ are conjugate to one another in regard to the conic Σ. The conic Ω is thus the director circle of Σ in regard to the points I, J; while Σ is the dual of the director circle of Ω in regard to the lines i and j.

It is seen that P, Q, Q' is a triad of points, on the conic Ω, of which two joins PQ, PQ' touch the conic Σ, while the third join QQ' passes through one point, O, of the self-polar triad common to Ω and Σ. Similarly, if the second tangent to Σ from Q' meet Ω in R, then RQ, QP, PQ' is a triad of three tangents of Σ, two of whose intersections, Q and P, lie on Ω, while the third, the intersection of RQ and PQ', lies on the line, HI, of the common self-polar triad of

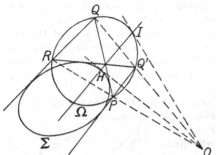

Ω and Σ. The tetrad P, Q, R, Q', of points of Ω, is such that the joins of these, taken in order, are tangent lines of Σ.

Conversely, if any two conics, Ω, Σ, be such that there exists a tetrad of points, P, Q, R, Q', of Ω, whose four joins, when the points are taken in order, are tangents of Σ, the diagonals PR and QQ' meet in one point of the common self-polar triad of the two conics; and there is then an infinite number of such tetrads of points. A sufficient condition for this relation is that the tangents of Σ, at two of its intersections with Ω, should meet in a point lying on Ω. Particular examples of two conics in this relation are, (1) Two circles Ω, Σ, of which Ω contains the centre of Σ, (2) A circle Ω passing through a pair of conjugate foci of Σ, (3) A circle Ω and a conic Σ which is the envelope of a chord PQ of Ω which subtends a right angle at a fixed point (not lying on Ω). In this last case, the middle point of P, Q describes a circle whose centre is the middle point of the two points constituted by O and the centre of Ω. (Cf. Ex. 31, Chap. I.)

CHAPTER III

THE EQUATION OF A LINE, AND OF A CONIC

Preliminary remarks. All the properties of conics so far given have been proved without the use of the symbols; and the same plan could be carried out to the end. For many of these properties it is, at present, more usual to give either a proof with the help of the symbols, or a proof by the metrical methods of what is called elementary geometry, in which, to a segment, or to an angle, a number is associated called its measure. We shall see that this metrical geometry can be regarded as an abbreviated use of algebraical symbols; it is, moreover, in the writer's view, in virtue of the system of symbols which it presupposes, founded on an arbitrary limitation of the geometrical possibilities. There is, therefore, logical interest in shewing that the symbols are not necessary; and in finding, if, for practical reasons, it is desired to reach the usual geometrical results, what are the narrowest geometrical limitations sufficient for this. But the employment of the symbols, by fixing the ideas, often renders a demonstration easier to follow. Moreover it happens that, since the time of Descartes, much labour has been devoted to elaborating systematic methods of algebraic computation; and, thus, methods which, from a geometrical point of view, are highly artificial, have come to be regarded as more natural.

The coordinates of a point in a plane. If A, B, C be three given points in a plane, not lying in line, of which the symbols are also represented by A, B, C, any other point in the plane has a symbol of the form $xA + yB + zC$, wherein x, y, z are algebraic symbols, of the system employed. These we suppose commutative in multiplication. If the point be specified for which the symbol is $A + B + C$, the symbols x, y, z are definite when P is assigned, save that they may be replaced, respectively, by mx, my, mz, where m is any algebraic symbol; thereby the symbols $x^{-1}y$, $x^{-1}z$ would be unaffected. The aggregate (x, y, z) are then called the coordinates of P, in regard to A, B, C, it being understood that (mx, my, mz) are equally the coordinates of P.

If we take any other three points, A', B', C', not lying in line, we may, similarly, specify the coordinates of all points of the plane in terms of these. In particular, if A, B, C be given in terms of these by the respective syzygies

$$A = a_1 A' + a_2 B' + a_3 C', \quad B = b_1 A' + b_2 B' + b_3 C',$$
$$C = c_1 A' + c_2 B' + c_3 C',$$

the symbol for P, with respect to A', B', C', becomes

$$x(a_1 A' + a_2 B' + a_3 C') + y(b_1 A' + b_2 B' + b_3 C') + z(c_1 A' + c_2 B' + c_3 C'),$$

or, say,
$$x'A' + y'B' + z'C',$$

where

$$x' = a_1 x + b_1 y + c_1 z, \quad y' = a_2 x + b_2 y + c_2 z, \quad z' = a_3 x + b_3 y + c_3 z.$$

The coordinates are thus changed by a linear transformation.

There can be no objection to agreeing to represent the relations involved, respectively, by

$$(A, B, C) = \begin{vmatrix} a_1, a_2, a_3 \\ b_1, b_2, b_3 \\ c_1, c_2, c_3 \end{vmatrix} (A', B', C'), \quad (x', y', z') = \begin{vmatrix} a_1, b_1, c_1 \\ a_2, b_2, c_2 \\ a_3, b_3, c_3 \end{vmatrix} (x, y, z);$$

herein the matrix of transformation by which we pass from the co-ordinates, x, y, z, to the new coordinates, x', y', z', is obtained from that by which we pass from the *new* points of reference, A', B', C', to the *original* points of reference, A, B, C, by the change, often called transposition, of rows into columns. This is often expressed by saying that (x, y, z) and (A, B, C) are transformed *contragrediently*.

The equation of a line. Let two points which determine a line be of respective symbols

$$P_1 = x_1 A + y_1 B + z_1 C, \quad P_2 = x_2 A + y_2 B + z_2 C;$$

any other point of the line is then of symbol $P_1 + \lambda P_2$, and has coordinates $x_1 + \lambda x_2$, $y_1 + \lambda y_2$, $z_1 + \lambda z_2$. Now consider the three symbols

$$u = y_1 z_2 - y_2 z_1, \quad v = z_1 x_2 - z_2 x_1, \quad w = x_1 y_2 - x_2 y_1;$$

we notice, first, that these are unaffected, save by multiplication with the same factor, if (x_1, y_1, z_1), (x_2, y_2, z_2) be, respectively, replaced by the coordinates, (x_1', y_1', z_1'), (x_2', y_2', z_2'), of any other two points of the line; thus, when the line is given, the symbols u, v, w are determined, save for a common multiplier; then we notice that, if (x, y, z) be the coordinates of any point whatever, of the line, we have

$$ux + vy + wz = 0;$$

finally, if u, v, w be given, any point whose coordinates (x, y, z) satisfy this equation, is a point of the line. For, (x_1, y_1, z_1), (x_2, y_2, z_2) being two points of the line, this equation, coupled with

$$ux_1 + vy_1 + wz_1 = 0, \quad ux_2 + vy_2 + wz_2 = 0,$$

enables us to infer that there are two symbols λ_1, λ_2 such that

$$x = \lambda_1 x_1 + \lambda_2 x_2, \quad y = \lambda_1 y_1 + \lambda_2 y_2, \quad z = \lambda_1 z_1 + \lambda_2 z_2.$$

The equation $\qquad ux + vy + wz = 0,$

wherein u, v, w are given (save for a common multiplier), and x, y, z have any values for which the equation is verified, being the relation satisfied by the coordinates of all points of the line, and by no other points, is called the *equation of the line*.

The coordinates of a line. The three symbols u, v, w, which, as we have seen, determine a line, are called the *coordinates of the line*. If m be any algebraic symbol, other than 0, the symbols mu, mv, mw are equally the coordinates of the line.

In particular, the coordinates of the fundamental points B, C, being, respectively, $(0, 1, 0)$ and $(0, 0, 1)$, the coordinates of the line BC, joining these, are $(1, 0, 0)$. If, then, we represent the *lines BC, CA, AB*, respectively, by the symbols a, b, c, we may regard the line whose coordinates are u, v, w as represented by the symbol

$$l = u\text{a} + v\text{b} + w\text{c}.$$

This is analogous to the representation of any point of the plane by means of the three fundamental points A, B, C. The analogy extends further. For we can shew that if l_1, l_2 be the symbols of any two lines, the symbol, formed in the same way, for any line through the intersection of these two, is of the form $pl_1 + ql_2$. To this end, let (x_0, y_0, z_0) be the coordinates of the point of intersection, O, of the lines, (x_1, y_1, z_1) of any other point of the first line, and (x_2, y_2, z_2) of any other point of the second line. Any line through O is determined by O and a point, say P, not lying on either of the two given lines; if we put

$$O = x_0 A + y_0 B + z_0 C, \quad P_1 = x_1 A + y_1 B + z_1 C, \quad P_2 = x_2 A + y_2 B + z_2 C,$$

we can suppose the symbol of P to be

$$P = \xi O + \eta P_1 + \zeta P_2;$$

thus, if P be $xA + yB + zC$, we have

$$x = \xi x_0 + \eta x_1 + \zeta x_2, \quad y = \xi y_0 + \eta y_1 + \zeta y_2, \quad z = \xi z_0 + \eta z_1 + \zeta z_2,$$

and these give

$$yz_0 - y_0 z = \eta (y_1 z_0 - y_0 z_1) + \zeta (y_2 z_0 - y_0 z_2),$$

with similar expressions for $zx_0 - z_0 x$ and $xy_0 - x_0 y$. Therefore, in terms of the coordinates of the two given lines, the coordinates of the line OP are

$$\eta u_1 + \zeta u_2, \quad \eta v_1 + \zeta v_2, \quad \eta w_1 + \zeta w_2,$$

and the symbol for this line is, therefore,

$$(\eta u_1 + \zeta u_2)\,\text{a} + (\eta v_1 + \zeta v_2)\,\text{b} + (\eta w_1 + \zeta w_2)\,\text{c},$$

or $\qquad\qquad\qquad \eta l_1 + \zeta l_2.$

The analogy between the expression of any point in terms of three points, and of any line in terms of three lines, is, thus, complete. And, as in the former case, so in the latter, when the symbols, a, b, c, of three fundamental lines are given, in order to specify without ambiguity the symbol of any other line, it is necessary to specify the symbol of some one fourth definite line beside a, b, c; we may, in particular, specify the line of which the symbol is a + b + c.

The equation of a point. We introduced the equation of a line as the equation satisfied by the coordinates of all points of the line. From what has just been said, we may, similarly, speak of the equation of a point, this being the equation satisfied by the coordinates of all the lines which contain this point. Clearly, if the point be of coordinates (x_0, y_0, z_0), the equation of the point will be

$$ux_0 + vy_0 + wz_0 = 0,$$

where u, v, w are variable line coordinates. For the sake of distinctness, the coordinates of a line are often called the *tangential* coordinates, and the equation of a point, its *tangential* equation.

And, just as the coordinates of the line joining the two points, whose coordinates are (x_1, y_1, z_1) and (x_2, y_2, z_2), are

$$u = y_1 z_2 - y_2 z_1, \quad v = z_1 x_2 - z_2 x_1, \quad w = x_1 y_2 - x_2 y_1,$$

so the coordinates of the point, which is the intersection of the lines of coordinates (u_1, v_1, w_1) and (u_2, v_2, w_2), are

$$x_0 = v_1 w_2 - v_2 w_1, \quad y_0 = w_1 u_2 - w_2 u_1, \quad z_0 = u_1 v_2 - u_2 v_1.$$

For, if (x_1, y_1, z_1) and (x_2, y_2, z_2) be the coordinates of two points, one on each line, the coordinates u_1, v_1, w_1 are the cofactors of x_2, y_2, z_2, respectively, in the determinant

$$\begin{vmatrix} x_0, & y_0, & z_0 \\ x_1, & y_1, & z_1 \\ x_2, & y_2, & z_2 \end{vmatrix},$$

and u_2, v_2, w_2 are, similarly, the cofactors of x_1, y_1, z_1, respectively. If we form the determinant whose elements are the minors of this determinant, the minors of the new determinant are known to be proportional to the elements of this one; this gives the statement made.

Thus, instead of regarding the symbols of the *points* of a plane as being fundamental, we might equally well have regarded the symbols of the *lines* as fundamental. Or, if we regard the *coordinates* as fundamental, we may regard that which we have from the first spoken of as the symbol of a point, as being, in fact, that expression, linear in undetermined line coordinates, which occurs above as the first member, or left side, of the equation of a point.

The equation of a conic. That the coordinates of any point of a line are connected by an equation may be regarded as a consequence of the fact that these coordinates are functions of a varying parameter, that is, of the forms

$$x = \theta x_1 + x_2, \quad y = \theta y_1 + y_2, \quad z = \theta z_1 + z_2,$$

wherein θ varies from point to point of the line. Now, we have shewn (Chap. I, above) that the symbol of any point of a non-degenerate conic is of the form

$$P = \theta^2 A + \theta B + C,$$

A, C being any two points of the conic, and B the pole of AC in regard to the conic; with reference to these points, the coordinates of any point of the conic are, then, $x = \theta^2$, $y = \theta$, $z = 1$. Thus every point of the conic is such that its coordinates satisfy the equation

$$xz = y^2.$$

With reference, however, to any three general points of the plane, A', B', C', which are not in line, the coordinates of a point of the conic, by what we have seen above, will be of the forms

$$x' = a_1 \theta^2 + b_1 \theta + c_1, \quad y' = a_2 \theta^2 + b_2 \theta + c_2, \quad z' = a_3 \theta^2 + b_3 \theta + c_3,$$

wherein the coefficients $a_1, b_1, \ldots, c_3$ remain the same, but θ varies from point to point of the conic. As we suppose that the conic does not degenerate into two straight lines, so that there is no relation of the form $Lx' + My' + Nz' = 0$, holding for all values of θ, these equations lead to equations

$$\theta^2 = A_1 x' + A_2 y' + A_3 z', \quad \theta = B_1 x' + B_2 y' + B_3 z', \quad 1 = C_1 x' + C_2 y' + C_3 z',$$

and, hence, to

$$(A_1 x' + A_2 y' + A_3 z')(C_1 x' + C_2 y' + C_3 z') = (B_1 x' + B_2 y' + B_3 z')^2.$$

Thus, referred to any three general points of the plane, the coordinates of all the points of a conic satisfy a homogeneous equation of the second order.

Conversely, any such equation, say

$$ax^2 + by^2 + cz^2 + 2fyz + 2gzx + 2hxy = 0,$$

connecting the coordinates x, y, z of a point, involves that the point lies on a definite conic. This we now prove:

(a) The quadric expression on the left side of this equation may be the product of two factors, both linear in x, y, z, being identically equal, say, to

$$(l_1 x + m_1 y + n_1 z)(l_2 x + m_2 y + n_2 z),$$

for all values of x, y, z. When this is so, by forming the partial differential coefficients in regard to x, y, z, we see that all the three linear functions of x, y, z given by

$$ax + hy + gz, \quad hx + by + fz, \quad gx + fy + cz$$

vanish for the same set of values of x, y, z, namely for that set of values which satisfies both the equations

$$l_1x + m_1y + n_1z = 0, \quad l_2x + m_2y + n_2z = 0;$$

for the moment we suppose that the lines of which these last are the equations are not the same. The simultaneous vanishing of the three linear functions requires, however, that the determinant

$$\Delta, = \begin{vmatrix} a, & h, & g \\ h, & b, & f \\ g, & f, & c \end{vmatrix},$$

should vanish. And, conversely, it will appear that it is easy to prove that when this determinant vanishes, the quadric expression, on the left side of the given equation, necessarily breaks up into the product of two linear factors. These two factors may be the same; but, then, there are other conditions connecting the coefficients $a, b, c, \dots$; in fact the minors, of two rows and columns, in Δ, also vanish. Also, it is true that if the given quadric equation is satisfied by the coordinates of every point of a line, say, the line of which the equation is $L = 0$, where L is linear in x, y, z, then the quadric expression on the left of the equation breaks up into the product of L and another linear factor. For, by ordinary division, we can write this quadric expression in such a form as

$$LM + A(y - mz)(y - nz),$$

where M is linear in x, y, z, and A, m, n are independent of x, y, z. This cannot be satisfied by every point for which $L = 0$ unless the coefficient A be zero.

(*b*) Now, suppose the determinant Δ is not zero. Let P_1, P_2 be any two points whose coordinates satisfy the given equation. From the fact that the given equation is of the second order, it can be shewn that every line drawn through P_1 contains another point, say P, whose coordinates also satisfy the given equation, so that there is associated to every line drawn through P_1, a definite line through P_2, joining this to P. If the coordinates of P_1 and P_2 be, respectively, (x_1, y_1, z_1) and (x_2, y_2, z_2), the equations of lines through P_1 and P_2, respectively, are of the forms

$$yz_1 - y_1z = \theta(xz_1 - x_1z), \quad yz_2 - y_2z = \phi(xz_2 - x_2z),$$

for proper values of θ and ϕ. By what we have seen, to every θ we can associate a definite ϕ, and conversely. Thus (cf. the Preliminary remarks to this volume, p. 8, above), the pencils $P_1(P)$ and $P_2(P)$ are related. Therefore P describes a conic in accordance with our original definition.

(*c*) We may reach this conclusion more directly, as follows: Let the quadric expression on the left side of the given equation be denoted by F, and the results of substituting therein, for x, y, z, respectively x_1, y_1, z_1 and x_2, y_2, z_2, by F_1 and F_2; let p, q, r denote, respectively, $ax + hy + gz$, $hx + by + fz$, $gx + fy + cz$, and p_1, q_1, r_1 what these become when x_1, y_1, z_1 are put for x, y, z, with, similarly, p_2, q_2, r_2, when x_2, y_2, z_2 are put for x, y, z; we then have

$$x_1 p_2 + y_1 q_2 + z_1 r_2 = x_2 p_1 + y_2 q_1 + z_2 r_1,$$

either of these being the same as

$$ax_1 x_2 + by_1 y_2 + cz_1 z_2 + f(y_1 z_2 + y_2 z_1) + g(z_1 x_2 + z_2 x_1) + h(x_1 y_2 + x_2 y_1),$$

which we denote by F_{12}. If we now put

$$X = xp_1 + yq_1 + zr_1, \quad Z = xp_2 + yq_2 + zr_2,$$
$$Y = x(y_1 z_2 - y_2 z_1) + y(z_1 x_2 - z_2 x_1) + z(x_1 y_2 - x_2 y_1),$$

and suppose the points (x_1, y_1, z_1), (x_2, y_2, z_2) to satisfy the given equation $F = 0$, so that both F_1 and F_2 vanish, we can shew that, for all values of x, y, z, the equation $F = 0$ is the same as

$$XZ - \mu Y^2 = 0,$$

where $\mu = \frac{1}{2}\Delta/F_{12}$. For, whatever x_1, y_1, z_1 and x_2, y_2, z_2 may be, we have, by multiplication of determinants, the product

$$\begin{vmatrix} x, & y, & z \\ x_1, & y_1, & z_1 \\ x_2, & y_2, & z_2 \end{vmatrix} \begin{vmatrix} x, & y, & z \\ x_1, & y_1, & z_1 \\ x_2, & y_2, & z_2 \end{vmatrix} \begin{vmatrix} a, & h, & g \\ h, & b, & f \\ g, & f, & c \end{vmatrix}$$

equal to

$$\begin{vmatrix} x, & y, & z \\ x_1, & y_1, & z_1 \\ x_2, & y_2, & z_2 \end{vmatrix} \begin{vmatrix} p, & q, & r \\ p_1, & q_1, & r_1 \\ p_2, & q_2, & r_2 \end{vmatrix},$$

and, hence, equal to

$$\begin{vmatrix} F, & X, & Z \\ X, & F_1, & F_{12} \\ Z, & F_{12}, & F_2 \end{vmatrix};$$

namely, we have the identity,

$$Y^2 \Delta = F(F_1 F_2 - F_{12}^2) + 2XZF_{12} - F_1 Z^2 - F_2 X^2.$$

When we substitute for x, y, z, in F, respectively, $x_1 + \lambda x_2, y_1 + \lambda y_2$, $z_1 + \lambda z_2$, the result is $F_1 + 2\lambda F_{12} + \lambda^2 F_2$; we are supposing $F_1 = 0$, $F_2 = 0$, and that the line joining (x_1, y_1, z_1) and (x_2, y_2, z_2) does not consist of points all satisfying $F = 0$; thus F_{12} does not vanish. The identity obtained thus leads to

$$F_{12}^2 F = 2F_{12}(XZ - \mu Y^2),$$

where $\mu = \frac{1}{2}\Delta/F_{12}$. Wherefore, the equation $F = 0$ is satisfied by all points (x, y, z) which, for any value of θ, satisfy the equations

$$X = \theta^2, \quad \mu Y = \theta, \quad \mu Z = 1;$$

thus the equation $F = 0$ belongs to a conic, provided the linear functions of x, y, z which we have denoted by X, Y, Z, are independent, that is, provided the lines $X = 0$, $Y = 0$, $Z = 0$ do not meet in a point; and, in fact, $X = 0$, $Y = 0$ intersect in the point (x_1, y_1, z_1), while $Z = 0$, $Y = 0$ meet in the distinct point (x_2, y_2, z_2).

A particular form of the equation of a conic. We have seen that, with reference to three points, consisting of any two points of a conic and the pole of the line joining them, the equation of the conic can be supposed to have the form $xz - y^2 = 0$, any point of the conic having coordinates of the form $(\theta^2, \theta, 1)$. Many results which hold for the general form of the equation can be well illustrated by this case.

(1) Evidently the line represented by the equation

$$x - y(\theta + \phi) + z\theta\phi = 0$$

contains both the points $(\theta^2, \theta, 1)$ and $(\phi^2, \phi, 1)$; this is, then, the equation of the chord of the conic. Or, a line represented by the equation $ax + by + cz = 0$ meets the conic in two points for which the values of the parameter θ are those given by the equation $a\theta^2 + b\theta + c = 0$. In particular the tangent of the conic at the point $x' = \theta^2, y' = \theta, z' = 1$ has the equation $x - 2y\theta + z\theta^2 = 0$; it is to be remarked that this is the same as $xz' + x'z - 2yy' = 0$, and, again, that this is the same as

$$\left(x' \frac{\partial}{\partial x} + y' \frac{\partial}{\partial y} + z' \frac{\partial}{\partial z}\right)(xz - y^2) = 0.$$

From this remark we can at once find the equation of the tangent at any point of the conic represented by the general equation

$$aX^2 + bY^2 + cZ^2 + 2fYZ + 2gZX + 2hXY = 0;$$

for, if X, Y, Z be any linear functions of x, y, z, say

$$X = a_1x + b_1y + c_1z, \quad Y = a_2x + b_2y + c_2z, \quad Z = a_3x + b_3y + c_3z,$$

and X', Y', Z' be precisely the same linear functions of x', y', z', it is easy to see that

$$X' \frac{\partial}{\partial X} + Y' \frac{\partial}{\partial Y} + Z' \frac{\partial}{\partial Z} = x' \frac{\partial}{\partial x} + y' \frac{\partial}{\partial y} + z' \frac{\partial}{\partial z};$$

now we have shewn that the general equation of the conic, say $F = 0$, in variable coordinates X, Y, Z, can, by choosing x, y, z to be suitable linear functions of these, be reduced to the form $xz - y^2 = 0$. Wherefore, the equation of the tangent, at a point of coordinates X', Y', Z', is

$$\left(X' \frac{\partial}{\partial X} + Y' \frac{\partial}{\partial Y} + Z' \frac{\partial}{\partial Z}\right) F = 0.$$

This, however, can be obtained, also, as follows: The point

(X', Y', Z') being upon the conic, and (X, Y, Z) being any other point, the point $(X' + \lambda X, Y' + \lambda Y, Z' + \lambda Z)$, on the line joining these, lies upon the conic, $F = 0$, if

$$\lambda^2 F + 2\lambda \left(X' \frac{\partial F}{\partial X} + Y' \frac{\partial F}{\partial Y} + Z' \frac{\partial F}{\partial Z} \right) + F' = 0,$$

where F' is what F becomes when X', Y', Z' are written for X, Y, Z, and is therefore zero. The equation will be satisfied by $\lambda = 0$ only, and the line joining (X', Y', Z') to (X, Y, Z) will meet the conic only at (X', Y', Z'), if, and only if, (X, Y, Z) satisfy the equation which we have given as the equation of the tangent at (X', Y', Z').

We can hence infer the equation of the polar of a point in regard to a conic. For the equation just obtained for the tangent is symmetrical in regard to (X, Y, Z) and (X', Y', Z'). Therefore, if the tangents of the conic at the points (x_1, y_1, z_1) and (x_2, y_2, z_2) intersect in the point (ξ, η, ζ), so that we have two equations, arising as expressing this fact, say

$$(x_1, y_1, z_1 ; \ \xi, \eta, \zeta) = 0, \quad (x_2, y_2, z_2 ; \ \xi, \eta, \zeta) = 0,$$

each linear in both the two sets of coordinates which it contains, and symmetrical in regard to these two sets, we can infer that the line expressed by the equation

$$(x, y, z ; \ \xi, \eta, \zeta) = 0,$$

wherein x, y, z are the coordinates of a varying point, passes through (x_1, y_1, z_1) and (x_2, y_2, z_2). It is thus the polar line of (ξ, η, ζ). Its equation is therefore

$$\left(\xi \frac{\partial}{\partial x} + \eta \frac{\partial}{\partial y} + \zeta \frac{\partial}{\partial z} \right) F = 0,$$

where now, in F, the coordinates x, y, z are written for X, Y, Z. The same result is obtained by regarding the polar as the locus of a point, (x, y, z), which is the harmonic conjugate of (ξ, η, ζ), in regard to the points in which the conic is met by a line through (ξ, η, ζ). For these intersections must then have coordinates of the respective forms $(x + \lambda \xi, y + \lambda \eta, z + \lambda \zeta)$ and $(x - \lambda \xi, y - \lambda \eta, z - \lambda \zeta)$; the quadratic equation for λ which expresses that one of these points satisfies the equation $F = 0$, must then be without the term in λ.

In particular, the polar of the point (ξ, η, ζ), in regard to the conic $xz - y^2 = 0$, is given by the equation $x\zeta + z\xi - 2y\eta = 0$.

With this form for the equation of the conic, the tangent line, whose equation, we have seen, is $x - 2y\theta + z\theta^2 = 0$, has coordinates $u = 1$, $v = -2\theta$, $w = \theta^2$. Wherefore the coordinates of all tangent lines of the conic are such as to satisfy the equation

$$4wu - v^2 = 0.$$

This equation is often called the tangential equation of the conic. In view of what has been said in regard to the point equation, and as to the dual nature of a conic, and from the similarity of the relations for the coordinates of a line and of a point, it is at once clear, (1) that for a conic whose equation is given in general form, the tangential equation must be homogeneously of the second order in the coordinates, u, v, w, of any tangent of the conic, (2) that any such quadric equation connecting u, v, w must be the tangential equation of a conic, becoming a point pair when the equation breaks into two linear factors. But we may prove directly that the condition for the line, whose equation is $ux + vy + wz = 0$, to touch the conic given by

$$ax^2 + by^2 + cz^2 + 2fyz + 2gzx + 2hxy = 0,$$

is
$$Au^2 + Bv^2 + Cw^2 + 2Fvw + 2Gwu + 2Huv = 0,$$

where A, B, C, F, G, H are respectively the cofactors of a, b, c, f, g, h in the determinant Δ, so that $A = bc - f^2$, $F = gh - af$, etc. And then, from the well-known determinant identities such as

$$BC - F^2 = a\Delta, \quad GH - AF = f\Delta,$$

we see that we can pass from the tangential equation to the point equation by the same rule as that by which the former equation is deduced from the latter. To prove the statement made, we express that the line $ux + vy + wz = 0$ is the tangent at some point, (x', y', z'), of the conic, so that this equation is of the form

$$x(ax' + hy' + gz') + y(hx' + by' + fz') + z(gx' + fy' + cz') = 0.$$

Whence we have

$$\frac{ax' + hy' + gz'}{u} = \frac{hx' + by' + fz'}{v} = \frac{gx' + fy' + cz'}{w},$$

which, however, lead to

$$\frac{x'}{Au + Hv + Gw} = \frac{y'}{Hu + Bv + Fw} = \frac{z'}{Gu + Fv + Cw};$$

these, when we express that $ux' + vy' + wz' = 0$, give the condition specified.

As examples, the reader may prove the simple facts that the conic whose equation is obtained by rationalising $(fx)^{\frac{1}{2}} + (gy)^{\frac{1}{2}} + (hz)^{\frac{1}{2}} = 0$ has for its tangential equation $fvw + gwu + huv = 0$, the tangent at any point (x', y', z') of the conic having the equation

$$x(f/x')^{\frac{1}{2}} + y(g/y')^{\frac{1}{2}} + z(h/z')^{\frac{1}{2}};$$

this conic touches the joins, BC, CA, AB, of the three points of reference for the coordinates. Dually, the conic, through $A, B, C,$

whose equation is $fyz + gzx + hxy = 0$, has a tangential equation obtainable from $(fu)^{\frac{1}{2}} + (gv)^{\frac{1}{2}} + (hw)^{\frac{1}{2}} = 0$, the tangent at any point (ξ, η, ζ) having the equation $\xi^{-2}fx + \eta^{-2}gy + \zeta^{-2}hz = 0$.

We return now again to the equation $xz - y^2 = 0$. Let X, Z be any points of the line AC which are harmonic conjugates in regard to A and C, say, of symbols $mA - C$, $mA + C$, respectively, in terms of the symbols of the points A and C; their coordinates will then be, referred to A, B, C, respectively $(m, 0, -1)$ and $(m, 0, 1)$. The equations of the lines BX, BZ will then be, respectively, $x + mz = 0$ and $x - mz = 0$. Now we have agreed that our system of symbols shall be such that there is a symbol i for which $i^2 = -1$ (Vol. I, p. 165); for this we use always the same sign; the other symbol satisfying the same equation will then be $-i$. This being understood, let

$$\xi = i(x - mz), \quad \zeta = (x + mz), \quad \eta = 2m^{\frac{1}{2}}y,$$

where $m^{\frac{1}{2}}$, as usual, is one of the symbols, t, for which $t^2 = m$. We then have, as another form of the equation $xz - y^2 = 0$, the equation

$$\xi^2 + \zeta^2 = \eta^2;$$

and we have, writing $m^{-\frac{1}{2}}$ for $(m^{\frac{1}{2}})^{-1}$,

$$xA + yB + zC = \tfrac{1}{2}(\zeta - i\xi)A + \tfrac{1}{2}m^{-\frac{1}{2}}\eta B + \frac{1}{2m}(\zeta + i\xi)C$$

$$= \frac{\xi}{2im}(mA - C) + \eta\frac{1}{2m^{\frac{1}{2}}}B + \frac{\zeta}{2m}(mA + C).$$

The triad, B, X, Z, is self-polar in regard to the conic. Conversely, if, in the general equation

$$ax^2 + by^2 + cz^2 + 2fyz + 2gzx + 2hxy = 0,$$

the coordinates are referred to a triad which is self-polar in regard to the conic, we must have $f = 0$, $g = 0$, $h = 0$. For the polar of the point $(1, 0, 0)$, which in general is of equation $ax + hy + gz = 0$, must then be expressed by $x = 0$, with a similar remark for the points $(0, 1, 0), (0, 0, 1)$. If, then, we put $\xi = xa^{\frac{1}{2}}$, $\eta = iyb^{\frac{1}{2}}$, $\zeta = zc^{\frac{1}{2}}$, the equation reduces to $\xi^2 + \zeta^2 = \eta^2$. The algebraic problem of reducing the general equation to this form, equivalent to the problem of finding a self-polar triad of the conic given by the general equation, is solved below (p. 141).

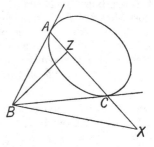

Equation of a circle. In the phraseology of the preceding chapter, with reference to the points A, C, taken as absolute points, the conic is a circle, whose centre is B; the lines BX, BZ are any two lines through its centre which are at right angles to one another in regard to A and C.

If, retaining X, Z as two of the points of reference, we take for the third point, instead of B, another point, B', the equation of the circle takes the form

$$(\xi' - a\eta)^2 + (\zeta' - c\eta)^2 = \eta^2.$$

In fact, writing (see p. 107)

$$X_1 = \frac{1}{2im}(mA - C), \quad Z_1 = \frac{1}{2m}(mA + C), \quad B_1 = \frac{1}{2m^{\frac{1}{2}}}B,$$

and taking B' so that

$$B' = B_1 - aX_1 - cZ_1,$$

the symbol of a general point found above, $\xi X_1 + \eta B_1 + \zeta Z_1$, is $(\xi + a\eta)X_1 + \eta B' + (\zeta + c\eta)Z_1$, and this we write in the form $\xi' X_1 + \eta B' + \zeta' Z_1$. The points of reference are now, any point B' of the plane, together with any two points X, Z, on the absolute line AC, which are such that $B'X, B'Z$ are at right angles in regard to A and C. The centre of the circle is now the point for which the coordinates ξ', ζ', η are, respectively, $a, c, 1$.

In further illustration of the point of view adopted in the last chapter, let us now find the equation of a general circle of a coaxial system, that is, of a conic through four given points. Denote these points by I, J, A, B, using I, J as the notation for the two absolute points. The points of intersection, S, H, respectively of AJ, BI and of AI, BJ, are (Chap. II, above, p. 66) the limiting points of the coaxial system of circles. Let AB, IJ meet in Y, also IJ, SH in X, and AB, SH in Z; the triad S, H, Y is self-polar in regard to all conics through I, J, A, B. We refer the circles to the points Z, X, Y, and suppose the symbols so chosen that $I = \frac{1}{2}(X + iY)$, $J = \frac{1}{2}(X - iY)$, as is possible since I, J are harmonic in regard to X and Y.

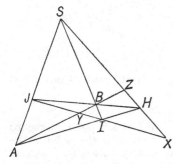

For any particular one of the circles, let the pole of the line IJ, which lies on XZ, be of symbol $Z + aX$; then, by what we have seen above, the symbol of any point of this circle is expressible, with a parameter θ, in the form

$$\theta^2 I + \theta m(Z + aX) + J,$$

where m is a fixed appropriate symbol. Denoting this by $xX + yY + zZ$, we have

$$x = \tfrac{1}{2}(\theta^2 + 1) + ma\theta, \quad y = \tfrac{1}{2}i(\theta^2 - 1), \quad z = m\theta,$$

which satisfy the equation

$$(x - az)^2 + y^2 = m^{-2}z^2,$$

other circles with the same centre having the same form of equation with a different value for m. But, if the points A, B be $Z \pm hY$, that is, be of coordinates $(0, \pm h, 1)$ with reference to X, Y, Z, this circle passes, by hypothesis, through A, B, and we have $m^{-2} = a^2 + h^2$. The circle may therefore be represented by

$$x^2 + y^2 - 2axz - h^2z^2 = 0,$$

and the other circles of the coaxial system, through A and B, have the same form of equation with a different value for a. The triad of reference for the coordinates, Z, X, Y, is determined by, (1) the line of centres of all the circles, which is the line XZ, (2) the common chord, YZ, of all the circles, (3) the absolute line, XY. The symbol h is such that the common points of the circles are of coordinates $(0, \pm h, 1)$.

The relation, to the points I, J, and the triad X, Y, Z, of the points S, H, is similar to that of A, B. But while A, B have symbols $Z \pm hY$, the symbols of S, H are $Z \pm kX$. Thus the general circle passing through the limiting points S, H has an equation

$$x^2 + y^2 - 2byz - k^2z^2 = 0,$$

where k is the same for all of these, but b varies from circle to circle. Here, as the points J, B, H, which may be supposed to have respective symbols $X - iY, Z + hY, Z + kX$, are in line, we may suppose $h = ik$; so that the general circle through the limiting points has an equation of the form

$$x^2 + y^2 - 2byz + h^2z^2 = 0.$$

We may now verify, what was proved in Chap. II (p. 66, above), that any one of these circles cuts any one of the coaxial circles at right angles. For this purpose we first obtain the condition that, at a point (x_0, y_0, z_0) where they intersect, two conics should cut at right angles, in regard to I and J, according to the definition we have given for this (Chap. II, p. 63). With regard to the points X, Y, Z, by which the coordinates are defined, we assume only that X, Y are on the absolute line, IJ, and are harmonic in regard to I and J; for these two absolute points, then, we take, as symbols, $X \pm iY$. If the equations of the two conics be, respectively, $\phi = 0$ and $\psi = 0$, and the values, at their common point (x_0, y_0, z_0), of the partial derivatives $\partial\phi/\partial x, \partial\phi/\partial y, \ldots, \partial\psi/\partial z$, be denoted, for brevity, by u, v, w, p, q, r, the tangents of these conics, at this point, are, we have seen, of equations $ux + vy + wz = 0$, and $px + qy + rz = 0$. These meet the absolute line, for which $z = 0$, respectively where $ux + vy = 0$ and $px + qy = 0$, that is, at the points whose symbols are $vX - uY$ and $qX - pY$. These symbols, however, are, respectively,

$$(v + iu)(X + iY) + (v - iu)(X - iY),$$
$$(q + ip)(X + iY) + (q - ip)(X - iY),$$

or $(v + iu)I + (v - iu)J, \quad (q + ip)I + (q - ip)J;$

the condition that the two tangents should be at right angles, in regard to I and J, is that these points should be harmonic conjugates in regard to I, J, namely, that

$$(v + iu)^{-1}(v - iu) + (q + ip)^{-1}(q - ip) = 0,$$

or $up + vq = 0;$

more fully this is, that, at the common point of the conics $\phi = 0$, $\psi = 0$, we should have

$$\frac{\partial \phi}{\partial x}\frac{\partial \psi}{\partial x} + \frac{\partial \phi}{\partial y}\frac{\partial \psi}{\partial y} = 0.$$

Applying this to the case of the circles whose equations are

$$x^2 + y^2 - 2axz - h^2z^2 = 0, \quad x^2 + y^2 - 2byz + h^2z^2 = 0,$$

it becomes $(x - az)x + y(y - bz) = 0;$

this is satisfied at each common point of the two circles, as we see by adding their equations.

Incidentally, we have seen that the condition that any two lines

$$ux + vy + wz = 0, \quad px + qy + rz = 0,$$

should be at right angles, in regard to I and J, is $up + vq = 0$.

Examples of the algebraic treatment of the theory of conics. The preceding explanations are probably sufficient to exhibit the application of the algebraic symbols to the theory of conics. We collect together now a succession of illustrations, with, in many cases, omission of detailed discussion. Most of the results obtained will be known to the reader of the preceding chapters; but there are some, of which the symbolic treatment is briefer than a purely geometrical treatment would be, which have not been referred to previously. The reader will be able to convince himself, by devising a geometrical method, that the symbols are not indispensable even for these. But there is one assumption which is made in some of the concluding examples to which reference should be made. It is assumed in these examples that the system of symbols employed is such that a rational algebraic equation of order n has n roots, and can be written as the product of n corresponding linear factors. Logically, the examples in which this assumption is made should have been deferred till after the considerations given in Chap. IV; it seemed more convenient to include them here.

Ex. 1. *Condition that a pair of points should be harmonic conjugates in regard to another pair.* On a line, determined by two points

O, U, we consider two points whose symbols are $O + \alpha U$, $O + \beta U$, where α, β are the values of x for which $ax^2 + 2hx + b = 0$, and also two points $O + \alpha' U$, $O + \beta' U$, where α', β' are the values of x for which $a'x^2 + 2h'x + b' = 0$. Prove that the condition, that the two latter points should be harmonic conjugates in regard to the two former, is that $(\alpha' - \alpha)(\alpha' - \beta)^{-1} + (\beta' - \alpha)(\beta' - \beta)^{-1} = 0$, and that this is the same as $ab' + a'b - 2hh' = 0$.

Ex. 2. A result in regard to involutions. If A, B and A', B' be two pairs of points, on a line, or on a conic, and X, Z be the double points of the involution which is determined by these two pairs, prove that the three pairs of points, A, A' ; B, B' ; X, Z, are in involution, as also are the three pairs, A, B' ; A', B ; X, Z.

Ex. 3. Representation of all the pairs of an involution. If, as in Ex. 1, two points, $A, = O + \alpha U$, and $B, = O + \beta U$, be determined by the roots α, β of the equation $ax^2 + 2hx + b = 0$, and, also, two points $A', = O + \alpha' U$, and $B', = O + \beta' U$, by the equation $a'x^2 + 2h'x + b' = 0$, then the two points, $O + \xi U$, $O + \eta U$, where ξ, η are the roots of the equation $\theta(ax^2 + 2hx + b) + a'x^2 + 2h'x + b' = 0$, for any symbol θ, are a pair of the involution determined by the two pairs A, B and A', B'. Further, every pair of this involution is so determined, with an appropriate value for θ.

Ex. 4. A general result for involutions. In the previous example, the points $O + \xi U$, $O + \eta U$ can be expressed by the points A and A'; let the expressions, for the symbols of the points, be

$$O + \xi U = m(A + \lambda A'), \quad O + \eta U = n(A + \mu A').$$

Prove, then, that $\lambda \mu^{-1} = (\xi - \alpha)(\xi - \alpha')^{-1}/(\eta - \alpha)(\eta - \alpha')^{-1}$; and, denoting this by ϵ, prove that

$$\frac{4\epsilon}{(1 + \epsilon)^2} = \frac{\theta[2(\Delta\Delta')^{\frac{1}{2}} - H]}{[\theta(\Delta)^{\frac{1}{2}} + (\Delta')^{\frac{1}{2}}]^2},$$

where $\Delta = h^2 - ab$, $\Delta' = h'^2 - a'b'$, $H = 2hh' - ab' - a'b$.

In particular, (Ex. 1), the points $O + \xi U$, $O + \eta U$ are harmonic conjugates of one another, in regard to A and A', when θ is such that

$$\theta^2(h^2 - ab) = h'^2 - a'b'.$$

Ex. 5. Condition that two circles, in general form, should cut at right angles. Shewing, from the text, that the equations of any two circles can be taken in the forms

$$x^2 + y^2 + 2gxz + 2fyz + cz^2 = 0, \quad x^2 + y^2 + 2g'xz + 2f'yz + c'z^2 = 0,$$

prove that the condition that they should cut at right angles is

$$2gg' + 2ff' = c + c'.$$

Ex. 6. The radical axis of two circles; the radical centre of three circles. If the equations of any two conics, referred to any triad of points, be $F = 0$, $F' = 0$, prove that the equation of any conic which passes through their four intersections is $\theta F + F' = 0$, for an appropriate symbol θ. In particular, shew that the radical axis of the two circles in the preceding example is (by $\theta = -1$)

$$2\,(g-g')\,x + 2\,(f-f')\,y + (c-c')\,z = 0.$$

Also that the three radical axes, for the pairs of any three circles, meet in a point. Shew, further (see Ex. 3), that the conics which pass through the common points of two given conics cut an arbitrary line in pairs of points belonging to the same involution.

Ex. 7. *The common tangents and centres of similitude of two circles.* Prove that the condition that the line whose equation is

$$ux + vy + wz = 0$$

should touch the circle whose equation is

$$(x - az)^2 + (y - bz)^2 = h^2z^2$$

is that $\qquad (au + bv + cw)^2 = h^2\,(u^2 + v^2).$

Hence, find the equations of the four tangents, common to this circle and the circle given by $(x - a'z)^2 + (y - b'z)^2 = h'^2z^2$. There are three point pairs such that every one of the four tangents passes through one or other of the two points of a pair; and the three lines, each containing the points of a pair, form a triad self-polar in regard to both the circles. Shew that these lines have for intersections the two limiting points of the two circles and a point on the absolute line, IJ. The centres of the two circles, in terms of the points of reference, X, Y, Z, have, for respective symbols, $C = aX + bY + Z$, and $C' = a'X + b'Y + Z$; shew that there are two points on the line joining the centres, with symbols $hC' \pm h'C$, through each of which pass two of the common tangents of the circles. These two points are called the centres of similitude of the two circles. It may be proved that the six centres of similitude of three circles lie, in threes, upon four lines.

The circle passing through the centres of similitude of two circles, with its centre on the line joining the centres of these, is of importance. Prove that the equation of this circle is $h^2F' - h'^2F = 0$, where $F = 0$, $F' = 0$ are the equations of the two circles, written as in this example.

Ex. 8. *Feuerbach's theorem. Apolar triads.* Let A, B, C be three points such that the joins BC, CA, AB touch a circle, say, Ω; and let D, E, F be the middle points, respectively, of B, C; of C, A; and of A, B. Then the circle D, E, F touches Ω. The definitions of a circle, and of middle points, are with reference to two absolute points I, J.

Let K be the centre of Ω, and take coordinates in regard to I, J, K, any point of the plane being $xI + zJ + yK$. The equation of Ω may then be supposed to be $xz - y^2 = 0$. Any point of this has, then, coordinates of the form $(\theta^2, \theta, 1)$; in particular, let the points of contact of BC, CA, AB be given, respectively, by the parameters α, β, γ, so that, for instance, the equation of BC is $x - 2y\alpha + z\alpha^2 = 0$. The coordinates of A are then easily found to be $\beta\gamma, \frac{1}{2}(\beta + \gamma), 1$, and the coordinates of D to be

$$\alpha(\alpha\beta + \alpha\gamma + 2\beta\gamma), \quad (\alpha + \beta)(\alpha + \gamma), \quad \beta + \gamma + 2\alpha.$$

Now write, for brevity,

$$p = \alpha + \beta + \gamma, \quad q = \beta\gamma + \gamma\alpha + \alpha\beta, \quad r = \alpha\beta\gamma, \quad \mu = q/p$$

and consider the conic whose equation is

$$(xz - y^2)(rp^{-2} - \mu) + y(x - 2y\mu + z\mu^2) = 0.$$

It can be verified that this equation is satisfied by the coordinates of D, E and F. The points common to this conic and $xz - y^2 = 0$ are the points of intersection of $xz - y^2 = 0$ with the two lines whose equations are $y = 0$, $x - 2y\mu + z\mu^2 = 0$. The former of these meets $xz - y^2 = 0$ in the points I and J; the latter meets it in two coincident points of coordinates $(\mu^2, \mu, 1)$. Thus the equation written down represents the circle D, E, F; and this touches Ω at the point $(\mu^2, \mu, 1)$.

As has appeared (Chap. I, Ex. 33; Chap. II, p. 89) the circle D, E, F is the locus of the centres of rectangular hyperbolas passing through A, B, C. This may be verified directly. The equation

$$\Sigma P(x - 2y\alpha + z\alpha^2)^{-1} = 0,$$

containing three terms with undetermined coefficients P, Q, R, evidently represents a conic through the three points A, B, C; by suitable choice of P, Q, R this can be taken to pass through two other arbitrary points; it is therefore the general conic through the points A, B, C. It is a rectangular hyperbola if its intersections with the line $y = 0$ are harmonic in regard to I and J; for this the equation, involving terms in x^2, xz, z^2, which gives its intersections with $y = 0$, must contain no term in xz; it is at once seen that the condition for this is $\Sigma P(\beta^2 + \gamma^2) = 0$. If we now form, by the rule given above, the equation for the polar of a point (x', y', z'), in regard to this conic, and express that this reduces to $y = 0$, (in which case (x', y', z') will be the centre of the conic), we obtain two further conditions for P, Q, R; in virtue of the first condition these become

$$\Sigma P[x' - y'(\beta + \gamma)] = 0 \quad \text{and} \quad \Sigma P\beta\gamma[y'(\beta + \gamma) - z'\beta\gamma] = 0.$$

Eliminating P, Q, R from the three conditions, we find that (x', y', z') must satisfy the equation

$$(xz - y^2)(r - pq) + y(xp^2 - 2ypq + zq^2) = 0,$$

which is that of the conic D, E, F.

The values of the parameters α, β, γ, for the points of contact of BC, CA, AB with the circle Ω, are the values of θ for which the polynomial $u = \theta^3 - p\theta^2 + q\theta - r$, vanishes. If we similarly take together the values of the parameter for the points I, J, of the circle Ω, and the point where it is touched by the circle D, E, F, these being, respectively, ∞, 0, μ, these may be said to be the roots of the polynomial $v = \theta^2 - \mu\theta$, regarded as a cubic polynomial. Now, for any two quadratic polynomials, say $a\theta^2 + 2h\theta + b$ and $a'\theta^2 + 2h'\theta + b'$, we have remarked (above, p. 111) an expression involving their coefficients, $ab' - 2hh' + ba'$, which is of geometrical interest. For any two cubic polynomials $a\theta^3 + 3b\theta^2 + 3c\theta + d$ and $a'\theta^3 + 3b'\theta^2 + 3c'\theta + d'$, there is a similar expression, which is also of geometrical interest, namely $ad' - 3bc' + 3cb' - da'$. Two cubics are said to be *apolar* to one another when this expression vanishes; this is the case for the two cubics, u, v, considered here. We may, therefore, say that the circle D, E, F touches the circle Ω at the point of Ω which is the apolar complement of I and J, with reference to the points where Ω is touched by BC, CA, AB. (See Ex. 26, below.)

Ex. 9. Continuation of the preceding example. The point where the nine points circle, of A, B, C, that is, the circle D, E, F, touches the inscribed circle Ω, may also be described as the fourth intersection of Ω with a particular conic, which we now explain, passing through the points where Ω touches BC, CA, AB. Let these points of contact be called P, Q, R ; the equation of the line QR is

$$x - y\,(\beta + \gamma) + z\beta\gamma = 0.$$

Denote this by $U = 0$. Similarly, let RP and PQ be $V = 0$ and $W = 0$. The conic represented by the equation

$$\left(x'\frac{\partial}{\partial x} + y'\frac{\partial}{\partial y} + z'\frac{\partial}{\partial z}\right) UVW = 0,$$

may be called the polar conic of the point (x', y', z') in regard to the degenerate cubic locus represented by $UVW = 0$. When (x', y', z') is the centre, $(0, 1, 0)$, of the circle Ω, this polar conic becomes $\Sigma\,(\beta + \gamma)\,VW = 0$. It may easily be verified that this conic passes through the point $(\mu^2, \mu, 1)$ of Ω, and that it passes through P, Q and R. Its tangent at P is the harmonic conjugate, in regard to PQ, PR, of the line joining P to the centre of the inscribed circle Ω, as is easy to verify, with a similar statement for Q and R; this suffices to identify the conic. Thus, the nine points circle of A, B, C touches the inscribed circle on the polar conic of the centre of the inscribed circle, taken in regard to the degenerate cubic formed of the three joins of the points of contact of the inscribed circle with BC, CA, AB.

Another conic which touches the inscribed circle Ω at the point

$(\mu^2, \mu, 1)$ is obtained as follows: Consider again the line drawn through P harmonic, in regard to PQ, PR, to the line joining P to the centre of the inscribed circle; and the similar lines through Q and R. A conic can be drawn touching these lines to have a focus at the centre of the inscribed circle. We may prove that this conic has the equation

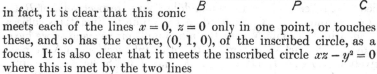

$$[x - 2y (\mu - rp^{-2}) + z\mu^2]^2$$
$$= 4xzr^2p^{-4};$$

in fact, it is clear that this conic meets each of the lines $x = 0$, $z = 0$ only in one point, or touches these, and so has the centre, $(0, 1, 0)$, of the inscribed circle, as a focus. It is also clear that it meets the inscribed circle $xz - y^2 = 0$ where this is met by the two lines

$$x - 2y\mu + z\mu^2 = 0, \quad x - 2y (\mu - 2rp^{-2}) + z\mu^2 = 0,$$

of which the former is the tangent at $(\mu^2, \mu, 1)$, and so touches the inscribed circle at this point. That it touches the line, spoken of, through P, whose equation is $W (\alpha + \gamma) + V (\alpha + \beta) = 0$, or $x (\alpha + p) - 2y (\alpha^2 + q) + z (\alpha q + r) = 0$, may be verified by shewing that the result of eliminating y between this equation and the equation of the conic, is a perfect square in x and z; or by first finding the tangential equation of the conic, which is

$$[2u (pq - r) + vp^2] [2w (pq - r) + vq^2] = v^2r^2,$$

and then verifying that this is satisfied by $u = \alpha + p$, $v = - 2 (\alpha^2 + q)$, $w = \alpha q + r$. The algebra is the same in both cases.

We may also give an algebraic proof of Feuerbach's theorem with coordinates which are referred to the points A, B, C. Without any loss of generality the equation of the circle touching BC, CA, AB may be taken to be

$$x^2 + y^2 + z^2 - 2yz - 2zx - 2xy = 0,$$

the absolute line, IJ, having the equation $lx + my + nz = 0$. Now consider the conic, say, N, expressed by the equation

$$2lmnC + fgh (lx + my + nz) (f^{-1}x + g^{-1}y + h^{-1}z) = 0,$$

where C is the expression $x^2 - 2yz + \ldots$ which occurs on the left side of the equation of the inscribed circle, and $f = m - n$, $g = n - l$, $h = l - m$. Its intersections with the circle expressed by $C = 0$ are the intersections of this with the line $lx + my + nz = 0$, and with the line $f^{-1}x + g^{-1}y + h^{-1}z = 0$. The former is the absolute line, so that the conic considered, N, is also a circle; the latter is in fact a tangent of the circle $C = 0$; thus the conic considered, N, is a circle touching the circle $C = 0$. To prove that the line $f^{-1}x + \ldots = 0$ touches the circle $C = 0$, we remark, more generally, that a conic whose equation is that obtained by rationalising the equation

$$(ax)^{\frac{1}{2}} + (by)^{\frac{1}{2}} + (cz)^{\frac{1}{2}} = 0,$$

which is the most general conic touching BC, CA, AB, has for its tangent at any point, (x', y', z'), the line given by

$$x\,(a/x')^{\frac{1}{2}} + y\,(b/y')^{\frac{1}{2}} + z\,(c/z')^{\frac{1}{2}} = 0,$$

and has the tangential equation $avw + bwu + cuv = 0$. Consider now the intersections of the conic N with BC, CA, AB; it is easily verified that three of these are the middle points, respectively, of B, C, of C, A, and of A, B, in regard to the absolute line. This suffices to identify the conic N with the nine points circle; and so we have proved Feuerbach's theorem. If the other intersections of N, respectively with BC, CA, AB, be D, E, F, and D' be the harmonic conjugate of D in regard to B and C, with a similar definition for E' and F', on CA and AB, it may be proved that D', E', F' lie on the line whose equation is

$$(x + y + z)\,(mn + nl + lm) - (l^2x + m^2y + n^2z) = 0.$$

Ex. 10. *Hamilton's modification of* Feuerbach's *theorem. Introductory results.*

The following results can be verified (cf. Chap. I, Ex. 33). Let A, B, C be any triad of points, not in line, and O, O' be two further points. Let AO, BO, CO meet BC, CA, AB, respectively, in D, E, F, and AO', BO', CO' meet BC, CA, AB, respectively, in D', E', F'; let EF, $E'F'$ meet in P, while FD, $F'D'$ meet in Q, and DE, $D'E'$ meet in R. Then the lines QR, RP, PQ, respectively, contain A, B, C, while the lines AP, BQ, CR meet in a point. Also, the lines DP, EQ, FR meet in a point, say, K, and the lines $D'P, E'Q, F'R$ meet in a point, say, K'. Further, the line KK' passes through the intersection of QR with BC, of RP with CA, and of PQ with AB.

A conic, say, S, can be drawn to touch BC, CA, AB, respectively, in D, E, F, and a conic, say, Ω, can be drawn to touch BC, CA, AB, respectively, in D', E', F'. And a conic, say, T, can be drawn to contain the six points D, D', E, E', F, F'. The points K, K' are the points of contact, respectively, with S and Ω, of their fourth common tangent, and these points lie on T. The triad P, Q, R is self-polar in regard to each of the three conics S, Ω, T. Further, if EF, FD, DE, respectively, meet BC, CA, AB in U, V, W, these points are in line, and the two intersections of this line with the conic Ω form, with K', a triad of points of Ω which is apolar (see below) with the triad D', E', F', of points thereon. Incidentally we have a construction, when three common tangents of two conics, and their points of contact, are given, for finding the points of contact of the remaining common tangent.

If the conics S and Ω be expressed, respectively, by the equations

$$(lx)^{\frac{1}{2}} + (my)^{\frac{1}{2}} + (nz)^{\frac{1}{2}} = 0, \quad (l'x)^{\frac{1}{2}} + (m'y)^{\frac{1}{2}} + (n'z)^{\frac{1}{2}} = 0,$$

and we denote $mn' - m'n$, $nl' - n'l$, $lm' - l'm$, respectively, by p, q, r, it will be found that the point P is $(-p, q, r)$, with similar forms for Q and R, and that K and K' are, respectively, $(p^2 l, q^2 m, r^2 n)$ and $(p^2 l', q^2 m', r^2 n')$. We can express the points of the conic Ω in terms of a parameter, θ, by the equations $l'x = \theta^2$, $m'y = (\theta - 1)^2$, $n'z = 1$, the points D', E', F' being then given, respectively, by $\theta = 0$, $\theta = 1$, $\theta = \infty$, and the point K' by $rn'\theta + pl' = 0$. The line UVW, whose equation is $lx + my + nz = 0$, meets Ω in points, I, J, given by the equation $\theta^2 (l/l') + (1 - \theta)^2 (m/m') + n/n' = 0$. For the three points I, J, K' the parameter values are then given by an equation of the form $a\theta^3 + \theta^2 - \theta + b = 0$, while the cubic equation for D', E', F' reduces to $\theta^2 - \theta$. This is what is meant by saying that the triad I, J, K' is apolar with the triad D', E', F' (see, above, Ex. 8). The equation of the conic T is

$$\Sigma ll' x^2 - \Sigma (mn' + m'n) yz = 0.$$

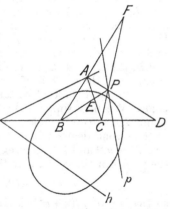

Ex. 11. *Hamilton's modification of Feuerbach's theorem.* Let the lines joining a point, P, of a conic, to three other points of the conic, A, B, C, meet BC, CA, AB, respectively, in D, E, F, so that D, E, F is a self-polar triad for the conic. We have remarked that the tangents at A, B, C meet BC, CA, AB, respectively, in the points of a line (Chap. I, Ex. 9), which we denote by h; we denote the tangent at P by p. We have already proved (Chap. I, Ex. 33) that if any conic having D, E, F for a self-polar triad be drawn, its four intersections with the lines p and h lie on a conic passing through A, B, C. We can take coordinates referred to D, E, F, in such a way that the tangents of the original conic at P, A, B, C have, respectively, the equations

$$x + y + z = 0, \quad -x + y + z = 0, \quad x - y + z = 0, \quad x + y - z = 0;$$

the original conic will then have an equation of the form

$$a^{-1}x^2 + b^{-1}y^2 + c^{-1}z^2 = 0,$$

wherein $a + b + c = 0$; any conic having D, E, F for self-polar triad will have an equation of the form $Ax^2 + By^2 + Cz^2 = 0$. The conic containing the four intersections of this with the lines p, h, and also the points A, B, C, will then have the equation

$$(Aa^2 + Bb^2 + Cc^2)(ax + by + cz)(x + y + z) = 4abc(Ax^2 + By^2 + Cz^2).$$

Conversely, if any conic be taken through A, B, C, meeting h and p, then a conic can be drawn through the four intersections, having D, E, F for a self-polar triad. If we take coordinates referred to A, B, C, and the equation of the original conic be put in the form $yz + zx + xy = 0$, while the conic taken through A, B, C has the equation $fyz + gzx + hxy = 0$, then the conic, through the four intersections of this with p and h, which has D, E, F for a self-polar triad, is

$$(f\eta\zeta + g\zeta\xi + h\xi\eta)(\xi^{-2}x + \eta^{-2}y + \zeta^{-2}z)(x + y + z)$$
$$+ fyz + gzx + hxy = 0,$$

where (ξ, η, ζ) are the coordinates of P.

Ex. 12. Conic referred to its centre and axes. The two lines drawn through the centre, C, of a conic, (the pole of the absolute line, IJ, in regard thereto), which are harmonic conjugates of one another both in regard to the two tangents which can be drawn to the conic from C, and in regard to CI, CJ, are, as we have previously said, called the axes of the conic. If these meet the line IJ in A and B, the triad A, B, C is self-polar, and, in reference to these, the equation of the conic is of the form $px^2 + qy^2 = z^2$. This fails if the conic is a parabola, that is, touches the line IJ, a possibility we exclude. We may suppose the coordinates x, y so chosen that the points I, J are on the lines $x \pm iy = 0$, so that, in any circle the terms of the second degree in x and y are $x^2 + y^2$. We replace p, q, respectively, by a^{-2} and b^{-2}. The expression in terms of a parameter, t, of the points of the conic, may then be taken to be

$$x = a(1 - t^2), \quad y = 2bt, \quad z = 1 + t^2,$$

so that $t = b^{-1}y/(z + a^{-1}x)$. It is usual to develop many results in this notation.

The equation

$$a^{-1}x(1 - t_1t_2) + b^{-1}y(t_1 + t_2) = z(1 + t_1t_2)$$

is at once verified to be that of the line joining the points t_1 and t_2; it gives the equation of the tangent by putting $t_2 = t_1$. The tangents to the curve from the points I, J pass through one of the two foci, S, H, given by $x = \pm (a^2 - b^2)^{\frac{1}{2}}$, $y = 0$, $z = 1$, and also through one of the other two foci, S', H', given by $x = 0$, $y = \pm (b^2 - a^2)^{\frac{1}{2}}$, $z = 1$. The normal of the conic at the point (ξ, η, ζ) has the equation

$$a^2\xi^{-1}x - b^2\eta^{-1}y = (a^2 - b^2)\zeta^{-1}z,$$

and this same equation, if we regard (x, y, z) as given, and ξ, η, ζ as variable, is that of the rectangular hyperbola of Apollonius, meeting the given conic at the feet of the four normals from (x, y, z), (p. 93).

If we consider the intersections of any circle with the given conic, substituting the above parametric expressions for the points of the conic in the equation of the circle, we immediately verify that any one of the three pairs of chords common to the conic and the circle meets the line IJ in two points which are harmonic in regard to A and B (a proposition often expressed by saying that these chords are equally inclined to one of the axes, cf. Chap. II, p. 79); and further, if t_1, t_2, t_3, t_4 be the parameters of the four intersections, and we assume (in illogical anticipation of the limitation of the symbols which is explained in the following Chapter) that the equation of the fourth order which these satisfy is capable of being written in the form $(t - t_1)(t - t_2)(t - t_3)(t - t_4) = 0$, we infer that $p_1 = p_3$, where p_r is the sum of the products of sets of r of t_1, t_2, t_3, t_4. In a similar way, from the equation for the Apollonius rectangular hyperbola, we infer that the parameters of the feet of the four normals which can be drawn to the conic from any point satisfy $1 - p_2 + p_4 = 0$. This last changes into the former if we replace t_4 by $- t_4^{-1}$. Thus if P_1, P_2, P_3, P_4 be the feet of the four normals, and Q_4 be the other point of the conic lying on the line $P_4 C$, it follows that P_1, P_2, P_3, Q_4 lie on a circle, as has already been proved (Chap. II, p, 93). If three points of the conic, with parameters t_1, t_2, t_3, be such that the normals thereat meet in a point, it can be verified that

$$\Sigma t_2 t_3 = \Sigma (t_2 t_3)^{-1}.$$

Lastly, if a line be drawn through one of the foci $(c, 0, 1)$, where $c^2 = a^2 - b^2$, at right angles to the tangent of the conic at any point, of equation $a^{-1}x(1 - t^2) + 2b^{-1}yt - z(1 + t^2) = 0$, this line will have the equation $2b^{-1}t(x - cz) - a^{-1}(1 - t^2)y = 0$, and will meet the tangent in the point whose coordinates are $x = a(1 - \lambda^2)$, $y = 2a\lambda$, $z = 1 + \lambda^2$, where $\lambda = (a - c)b^{-1}t$, this being a point of the circle $x^2 + y^2 = a^2 z^2$. This has been called the auxiliary circle (Chap. II, p. 85).

Ex. 13. *Confocal conics.* The points I, J, S, H, S', H', in the preceding example, having, respectively, the coordinates

$$(1, i, 0), \quad (1, -i, 0), \quad (c, 0, 1), \quad (-c, 0, 1), \quad (0, ic, 1), \quad (0, -ic, 1),$$

where $c^2 = a^2 - b^2$, any two conics touching the four lines $IH'S$, $JS'S$, $IS'H$, $JH'H$, may be taken to have equations

$$a^{-2}x^2 + b^{-2}y^2 = z^2, \quad (a^2 + \lambda)^{-1}x^2 + (b^2 + \lambda)^{-1}y^2 = z^2 \,;$$

these are *confocals*, in regard to I and J. The condition that any line, of equation $lx + my + nz = 0$, should touch the second of these conics is $a^2 b^2 + b^2 m^2 - n^2 - \lambda(m^2 + n^2) = 0$; the tangential equation

of the first conic is $a^2b^2 + b^2m^2 - n^2 = 0$, and the tangential equation
of the point pair I, J is $l^2 + m^2 = 0$. We have previously remarked
(p. 112, above) that if $f = 0$, $f' = 0$ be the point equations of any two
conics, the equation of the general conic through their four intersections is $\kappa f + f' = 0$, for a proper parameter κ; thus the tangential
equation of a confocal conic appears, as it should, as that of the
general conic touching the tangents from the points I, J to any one
conic of the system. It is easy to verify the properties of a system
of confocal conics, many of which have already been referred to
(pp. 79, 90, 93), by means of these equations. In particular, the
polar of a point (ξ, η, ζ), in regard to the conic
$$(a^2 + \lambda)^{-1}x^2 + \ldots - z^2 = 0,$$
has the equation
$$(a^2 + \lambda)^{-1}\xi x + (b^2 + \lambda)^{-1}\eta y - \zeta z = 0,$$
which we may write in the form $\lambda^2 Z + \lambda Y + X = 0$, where

$$Z = \zeta z, \quad Y = (a^2 + b^2)\zeta z - \xi x - \eta y, \quad X = a^2 b^2 \zeta z - a^2 \eta y - b^2 \xi x.$$

By what we have previously seen, this is the equation of the tangent
of the conic whose equation is $4XZ - Y^2 = 0$. It can be verified that
this is the same as is obtained by rationalising the equation
$$(\xi x)^{\frac{1}{2}} + (-\eta y)^{\frac{1}{2}} + c(\zeta z)^{\frac{1}{2}} = 0.$$
This conic touches the line $x = 0$ (for $\lambda = -a^2$), the line $y = 0$ (for
$\lambda = -b^2$), and the line $z = 0$ (for $\lambda = \infty$); it is thus a parabola, in
regard to I and J, the point C, at which two perpendicular tangents
meet, being on the directrix. The point (ξ, η, ζ) is equally on the
directrix; for there are two of the confocals passing through this
point, those namely for which λ has either of the values for which
$(a^2 + \lambda)^{-1}\xi^2 + \ldots - \zeta^2 = 0$, and these cut at right angles; for the tangents are $(a^2 + \lambda_1)^{-1}\xi x + \ldots - \zeta z = 0$ and $(a^2 + \lambda_2)^{-1}\xi x + \ldots - \zeta z = 0$,
which are at right angles (p. 110, above), if
$$(a^2 + \lambda_1)^{-1}(a^2 + \lambda_2)^{-1}\xi^2 + (b^2 + \lambda_1)^{-1}(b^2 + \lambda_2)^{-1}\eta^2 = 0,$$
an equation obtainable from
$$(a^2 + \lambda_1)^{-1}\xi^2 + \ldots - \zeta^2 = 0 \quad \text{and} \quad (a^2 + \lambda_2)^{-1}\xi^2 + \ldots - \zeta^2 = 0;$$
and these tangents are the polars of (ξ, η, ζ) in regard to these two
confocals. The focus of the parabola is the intersection of two of
its tangents passing respectively through I and J; the polar
$(a^2 + \lambda)^{-1}\xi x + \ldots - \zeta z = 0$ passes through $(1, i, 0)$ for the value of λ
given by $(a^2 + \lambda)^{-1}\xi + (b^2 + \lambda)^{-1}i\eta = 0$; and we similarly find the
polar passing through $(1, -i, 0)$. The intersection of these polars
gives the focus, which is found to be $(c^2\xi\zeta, -c^2\eta\zeta, \xi^2 + \eta^2)$. It is
easy to verify that the general conic passing through the four foci
has the equation $x^2 - y^2 - c^2 z^2 + 2hxy = 0$, where h is variable, and

hence that the focus of the parabola is conjugate to (ξ, η, ζ) in regard to all these conics. Further we may verify that the polar reciprocal of the parabola in regard to $a^{-2}x^2 + b^{-2}y^2 - z^2 = 0$, is the Apollonius rectangular hyperbola containing the feet of the normals of this conic which pass through (ξ, η, ζ); for, in order that the polar of a point (x', y', z'), in regard to the conic $a^{-2}x^2 + b^{-2}y^2 - z^2 = 0$, may coincide with a tangent of the parabola whose equation is

$$(a^2 + \lambda)^{-1}\xi x + (b^2 + \lambda)^{-1}\eta y - \zeta z = 0,$$

it is necessary and sufficient that

$$a^{-2}x'/(a^2 + \lambda)^{-1}\xi = b^{-2}y'/(b^2 + \lambda)^{-1}\eta = z'/\zeta;$$

this can be satisfied by a proper value of λ if

$$a^2\xi (x')^{-1} - b^2\eta (y')^{-1} = (a^2 - b^2)\zeta (z')^{-1}.$$

Ex. 14. Conics derived from a conic intersecting the joins of three points. If a conic, S, which meets the lines BC, CA, AB, respectively, in D, D'; E, E'; F, F', have the equation

$$ax^2 + by^2 + cz^2 + 2fyz + 2gzx + 2hxy = 0,$$

the lines AD, AD' are given by $by^2 + cz^2 + 2fyz = 0$. The coordinates of either of these lines, whose equation may be written $vy + wz = 0$, are then such that, $b^{-1}v^2 + c^{-1}w^2 - 2b^{-1}c^{-1}fvw = 0$. The six lines $AD, AD', BE, BE', CF, CF'$ are thus tangents of the conic, Σ', whose tangential equation is

$$bcu^2 + cav^2 + abw^2 - 2afvw - 2bgwu - 2chuv = 0,$$

or, say, $\Sigma' = 0$. If $\Sigma = 0$ be the tangential equation of the conic S, so that Σ is the sum of three pairs of terms such as

$$(bc - f^2)u^2 + 2(gh - af)vw,$$

we see that $\Sigma' - \Sigma$ is the sum of three pairs of terms such as $f^2u^2 - 2ghvw$. We have remarked that a conic whose equation is obtained by rationalising the equation $(lx)^{\frac{1}{2}} + (my)^{\frac{1}{2}} + (nx)^{\frac{1}{2}} = 0$ touches BC, CA, AB; similarly a conic whose tangential equation, $\Omega = 0$, is obtained by rationalising $(fu)^{\frac{1}{2}} + (gv)^{\frac{1}{2}} + (hw)^{\frac{1}{2}} = 0$, is one which passes through A, B, C; its point equation is, in fact, $fyz + gzx + hxy = 0$. In the case in hand we have $\Sigma' - \Sigma = \Omega$. The four common tangents of the conics S and Σ' are thus touched by the conic Ω, which passes through A, B, C. Similarly, by taking the point equation of the conic Σ', we see that there is a conic passing through the four common points of S and Σ' which touches BC, CA, AB, its point equation being $(afx)^{\frac{1}{2}} + (bgy)^{\frac{1}{2}} + (chz)^{\frac{1}{2}} = 0$.

A particular case of the preceding is when S is such that Σ' degenerates into a point pair, so that the lines AD, BE, CF meet in a point, say, O, and AD', BE', CF' also meet in a point, say, O'.

Then the points, on BC, CA, AB, which are the respective harmonic conjugates of D, E, F with regard to B, C; C, A; A, B, lie on a line, say, l; and the points, on BC, CA, AB, similarly derived from D', E', F', also lie on a line, say, l'. In general, these derived six points lie on the conic whose equation is

$$ax^2 + by^2 + cz^2 - 2fyz - 2gzx - 2hxy = 0,$$

as is easy to see; in the case now considered, this conic breaks into the two lines, l and l'. When this is so, it follows, from what has been said, that there is a conic through A, B, C which touches the tangents drawn to S from O and O'. From the equations it also follows easily that this conic passes through the intersections of S with the lines l and l'; and also, through these intersections there passes a conic in regard to which A, B, C are a self-polar triad, the lines l and l' being the polars, respectively, of O' and O, in regard to this. The point equation of a conic, given in tangential form, which breaks into a point pair, represents the line joining these points, taken twice over, as is easy to see. Thus, in the case now considered, there is a conic touching BC, CA, AB which touches S at its two intersections with the line OO'; and, further, the tangents of the conics, at their two points of contact, meet in the intersection of the lines l, l'. In the present case, we may regard S as the nine points circle of the triad, A, B, C; we have only to take for absolute points the intersections of S with the line l; it can be proved that these are conjugate points in regard to all conics through the four points A, B, C, O', so that, for instance, AO' is at right angles to BC in reference to these points. Thus the results here stated are equivalent to saying that the two points O, O', respectively the intersection of the lines from A, B, C to the middle points of B, C, of C, A, and of A, B, and the orthocentre of A, B, C, are the centres of similitude of the nine points circle and the circle A, B, C, the radical axis of these circles being the line l', derived from the orthocentre, O', by the rule given; further that there is a circle, coaxial with these two circles, with centre at the orthocentre, for which A, B, C is a self-polar triad, and for which O is the pole of the radical axis spoken of; and, also, that there is a conic touching BC, CA, AB, which touches the nine points circle at two points of the line OO' whereat the tangents are parallel (with regard to the absolute line l) to this radical axis. (See also Chap. i, Ex. 29.)

It would seem to be worth while to speak of the conic whose equation, referred to three points A, B, C, is

$$ax^2 - 2fyz + by^2 - 2gzx + cz^2 - 2hxy = 0,$$

as the harmonically conjugate of the conic whose equation is

$$ax^2 + 2fyz + by^2 + 2gzx + cz^2 + 2hxy = 0.$$

If the latter meet BC in P_1 and P_2, the former meets BC in two points, P_1', P_2', which are the harmonic conjugates, in regard to B and C, respectively, of P_1 and P_2, with a similar statement for CA and AB. And, further, to speak of the latter conic as being bipunctual, in regard to A, B, C, when the six lines such as AP_1, AP_2, meet in two points. The condition for this is that the harmonically conjugate conic break up into two lines.

If X, Y, Z be a self-polar triad in regard to a conic, and O be any point, and XO, YO, ZO meet YZ, ZX, XY, respectively, in A, B, C, then the conic is bipunctual in regard to A, B, C. For if we suppose O to be $(1, 1, 1)$, relatively to X, Y, Z, and the equation of the conic relatively to these be $ax^2 + by^2 + cz^2 = 0$, by taking

$$\xi = \tfrac{1}{2}(-x+y+z), \quad \eta = \tfrac{1}{2}(x-y+z), \quad \zeta = \tfrac{1}{2}(x+y-z),$$

we have, for the equation in regard to A, B, C, the form

$$a(\eta + \zeta)^2 + b(\zeta + \xi)^2 + c(\xi + \eta)^2 = 0.$$

The harmonically conjugate conic, in regard to A, B, C, is then

$$a(\eta - \zeta)^2 + b(\zeta - \xi)^2 + c(\xi - \eta)^2 = 0,$$

which is evidently a line pair.

Conversely, if a conic be bipunctual in regard to A, B, C, it is possible, and in only one way, to draw lines, YZ, ZX, XY, passing, respectively, through A, B, C, such that the triad X, Y, Z, formed by their intersections, is self-polar in regard to the conic; and the lines AX, BY, CZ meet in a point. For, if, referred to A, B, C, the conic be

$$ll'\xi^2 + mm'\eta^2 + nn'\zeta^2 - (mn' + m'n)\,\eta\zeta - (nl' + n'l)\,\zeta\xi - (lm' + l'm)\,\xi\eta = 0,$$

and we attempt to write this equation in the form

$$P(\eta + p\xi)^2 + Q(\zeta + q\xi)^2 + R(\xi + r\eta)^2 = 0,$$

we find

$$p = (nl' - n'l)(lm' - l'm)^{-1}, \quad -2Pp = mn' + m'n,$$

with similar values for q, Q, r, R, (so that $pqr = 1$).

We may similarly consider a conic such that the tangents to it from A, B, C meet BC, CA, AB in points lying, in threes, upon two lines.

Ex. 15. *Two triads reciprocal in regard to a conic are in perspective.* If P, Q, R be the poles, with regard to any conic whose tangential equation referred to A, B, C is

$$Au^2 + Bv^2 + Cw^2 + 2Fvw + 2Gwu + 2Huv = 0,$$

respectively of BC, CA, AB, then the lines AP, BQ, CR meet in a point (Ex. 10, Chap. I). For P, Q, R have the respective coordinates (A, H, G), (H, B, F), (G, F, C), and the lines AP, BQ, CR meet in the point given by

$$Fx = Gy = Hz.$$

Dually, the polars of A, B, C meet BC, CA, AB, respectively, in three points lying in line. If the equation of the conic be

$$ax^2 + \ldots + 2fyz + \ldots = 0,$$

the equation of this line is $f^{-1}x + g^{-1}y + h^{-1}z = 0$.

Ex. 16. *The Hessian line of three points of a conic.* If the tangents at three points, A, B, C, of a conic, meet BC, CA, AB, respectively in D, E, F, the points D, E, F are in line (Chap. I, Ex. 9). If the points of the conic be expressed by a parameter, and the values of this for the points A, B, C be the values of x/y for which $ax^3 + 3bx^2y + 3cxy^2 + dy^3 = 0$, the values of the parameter at the two points, H, K, at which the line D, E, F meets the conic, are the two values of x/y for which

$$\frac{\partial^2 F}{\partial x^2} \cdot \frac{\partial^2 F}{\partial y^2} - \left(\frac{\partial^2 F}{\partial x \partial y}\right)^2 = 0,$$

where $F = 0$ is the equation above (cf. Vol. I, p. 174). Further, if the left side of the last quadratic be multiplied by $x + my$, *where m is arbitrary*, and the result be written $a'x^3 + 3b'x^2y + 3c'xy^2 + d'y^3 = 0$, it may be proved that $ad' - 3bc' + 3cb' - da' = 0$. In other words, any three points of a conic are apolar with the Hessian points of these taken with a third arbitrary point. If we take the equation of the conic in the form $fyz + gzx + hxy = 0$, the points (x, y, z) can be expressed in terms of a parameter by $x = -f\theta^{-1}$, $y = g(\theta - 1)^{-1}$, $z = h$, so that A, B, C correspond, respectively, to $\theta = 0, 1, \infty$; the points H, K are then given by $\theta^2 - \theta + 1 = 0$, and it is easy to verify the relation in question for the two cubics

$$x^2y - xy^2 = 0, \quad (x + my)(x^2 - xy + y^2) = 0.$$

The relation is unaffected if, instead of θ, we use a parameter ϕ connected with θ by an equation $A\theta\phi + B\theta + C\phi + D = 0$, wherein A, B, C, D are constants. Similar remarks hold for the Hessian point of three given tangents of a conic.

Ex. 17. *The director circle of a conic, and generalisations.* It can be proved that the equation of the two tangents drawn from a point (x', y', z') to touch the conic $U = 0$ is $UU' - T^2 = 0$, where U' is what U becomes when x', y', z' are written in place of x, y, z; and T is one half of what is obtained by operating upon U by

$$x'\partial/\partial x + y'\partial/\partial y + z'\partial/\partial z,$$

and is symmetrical in regard to x, y, z and x', y', z'. This result follows by considering that the points, with coordinates of the forms $(x + \theta x', y + \theta y', z + \theta z')$, on the line joining (x, y, z) to (x', y', z'), which lie on the conic, are given by $U + 2\theta T + \theta^2 U' = 0$.

Further, if these tangents meet the line joining the points, I, J, which have coordinates $(1, \pm i, 0)$, in two points harmonic in regard to these, that is, if the tangents be at right angles, then (x', y', z') lies on the locus whose equation is

$$(a + b)\, U = (ax + hy + gz)^2 + (hx + by + fz)^2,$$

or
$$C\,(x^2 + y^2) - 2Gxz - 2Fyz + (A + B)\, z^2 = 0\,;$$

here U is $ax^2 + \ldots + 2fyz + \ldots$, and $A = bc - f^2$, $F = gh - af$, etc. This gives the equation of the director circle in regard to I and J.

More generally, the locus of the point of intersection of two lines, conjugate to one another in regard to the conic, drawn respectively through the points (x_1, y_1, z_1), (x_2, y_2, z_2), is $U T_{12} = T_1 T_2$, where T_1 and T_2 are the forms of T when (x', y', z') are replaced, respectively, by (x_1, y_1, z_1) and (x_2, y_2, z_2), while T_{12} is obtained from T by replacing (x, y, z) and (x', y', z'), respectively, by (x_1, y_1, z_1) and (x_2, y_2, z_2). This locus becomes the director circle when the points (x_1, y_1, z_1), (x_2, y_2, z_2) are the points $(1, \pm i, 0)$.

This however may be generalised; we may seek the locus of a point from which the tangents to a conic are conjugate in regard, not to a point pair but, to another general conic, that is, the locus of a point from which the pair of tangents to one conic are harmonic conjugates of one another in regard to the tangents from that point to another conic. The locus is still a conic; and it is obvious by the definition that this contains the eight points of contact with the two conics of their four common tangents. If the tangential equation of one of the conics be $Au^2 + \ldots + 2Fvw + \ldots = 0$, that of the other being of the same form with $A', \ldots, F', \ldots$ in place of $A, \ldots, F, \ldots$, the pair of tangents to the former from the point (ξ, η, ζ) is given by

$$A\,(y\zeta - \eta z)^2 + \ldots + 2F\,(z\xi - \zeta x)\,(x\eta - \xi y) + \ldots = 0,$$

and these meet $z = 0$ in two points given, say, by $\mathrm{a}x^2 + 2\mathrm{h}xy + \mathrm{b}y^2 = 0$, where

$$\mathrm{a} = B\zeta^2 - 2F\zeta\eta + C\eta^2, \quad \mathrm{h} = -H\zeta^2 + F\xi\zeta + G\eta\zeta - C\xi\eta,$$
$$\mathrm{b} = A\zeta^2 - 2G\zeta\xi + C\xi^2\,;$$

the corresponding values for the second conic being denoted by $\mathrm{a}', \mathrm{h}', \mathrm{b}'$, we require to express that $\mathrm{a}\mathrm{b}' - 2\mathrm{h}\mathrm{h}' + \mathrm{b}\mathrm{a}' = 0$. On reduction this shews that (ξ, η, ζ) describes the conic whose equation is

$$(BC' + B'C - 2FF')\, x^2 + \ldots + 2\,(GH' + G'H - AF' - A'F)\, yz + \ldots = 0.$$

Dually, a line whose intersections with one conic,

$$ax^2 + \ldots + 2fyz + \ldots = 0,$$

are harmonic in regard to its intersections with another such conic, envelopes the conic whose tangential equation is

$$(bc' + b'c - 2ff')\, u^2 + \ldots + 2\,(gh' + g'h - af' - a'f)\, vw + \ldots = 0\,;$$

this conic evidently touches the eight tangents of the two conics at their common points.

For the former case, of two conics given tangentially, we may, by taking a pair of the intersections of their common tangents as the absolute points, I, J, regard the two conics as belonging to a system of confocals, their equations being

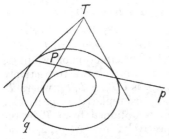

$$a^{-2}x^2 + b^{-2}y^2 - z^2 = 0,$$
$$(a^2 + \lambda)^{-1}x^2 + (b^2 + \lambda)^{-1}y^2 - z^2 = 0.$$

The tangents from a point, P, to the former conic, say, p and q, will be conjugate in regard to the second conic, if q passes through the pole, T, of the line p in regard to the second conic. The locus of P, with these coordinates, is found to be

$$2a^2b^2\left[a^{-2}x^2 + b^{-2}y^2 - z^2\right] + \lambda\left[x^2 + y^2 - (a^2 + b^2)z^2\right] = 0 ;$$

it meets either of the two conics in the four points where this is met by its director circle.

For the latter case, of two conics given in point coordinates, we may regard both the conics as circles. A line, l, will meet the circles harmonically if it join a point P, of one circle, to one of the points, Q, where this circle is met by the polar of P taken in regard to the other circle. The circles being supposed to have equations

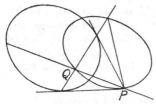

$$x^2 + y^2 + 2gxz + cz^2 = 0, \quad x^2 + y^2 + 2g'xz + cz^2 = 0,$$

as we have shewn to be possible (above, p. 109), the envelope is found to be given by

$$(u^2 + v^2)(c - gg') + (gu - w)(g'u - w) = 0.$$

The equation $u^2 + v^2 = 0$ represents the absolute points I, J, and the equation $(gu - w)(g'u - w) = 0$ represents the centres of the two circles. The envelope has the centres of the two circles for foci, and touches the tangents of the circles drawn at their intersections. A particular case is that in which the two circles cut at right angles ; then $c = gg'$, and the envelope reduces to the point pair given by the centres of the circles. We have previously seen (Chap. ii, p. 66) that when two circles cut at right angles, any line through the centre of either is divided harmonically by the circles.

A particular case is when one of the two conics is a line pair. For example, if the conic be the circle $x^2 + y^2 - (a^2 + b^2)z^2 = 0$, and the line pair be $a^{-2}x^2 + b^{-2}y^2 = 0$, the envelope is found to be $a^2u^2 + b^2v^2 - w^2 = 0$, which is the tangential equation of the conic $a^{-2}x^2 + b^{-2}y^2 - z^2 = 0$. Thus any tangent of a conic meets the director circle in two points which are harmonic in regard to the points in which the tangent meets the asymptotes of the conic.

The director circle, S, of a conic Σ, has been shewn (Chap. II, p. 96) to be such that sets of four points can be taken thereon whose joins, in order, are four tangents of the conic. It may be proved that for two conics $S = 0$, $\Sigma = 0$, where

$$S = ax^2 + by^2 + cz^2, \quad \Sigma = Au^2 + Bv^2 + Cw^2,$$

the condition for this possibility is that the sum of two of the quantities aA, bB, cC should be equal to the third.

Ex. 18. *Envelope of a line joining points of two conics which are conjugate in regard to a third.* We now prove the following theorem, which will be found to include many particular results: let

$$S_0 = x^2 + y^2 + z^2, \qquad \Sigma_0 = u^2 + v^2 + w^2,$$
$$S_1 = ax^2 + by^2 + cz^2, \quad \Sigma_1 = a^{-1}u^2 + b^{-1}v^2 + c^{-1}w^2,$$

so that $\Sigma_0 = 0$ is the tangential equation of the conic $S_0 = 0$, and $\Sigma_1 = 0$ is the tangential equation of the conic $S_1 = 0$. In general, any two conics have four distinct intersections, and their equations can be reduced to these forms. Any conic touching the four common tangents of Σ_0 and Σ_1 has a tangential equation $\Sigma_0 - \theta^2\Sigma_1 = 0$. We prove that this conic is the envelope of a line joining two points, lying respectively on the conics S_0 and S_1, say P and Z, which are

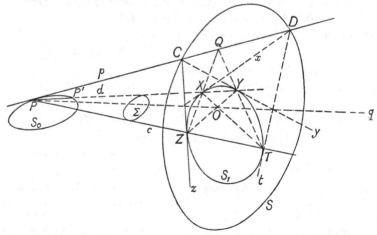

conjugate to one another in regard to the conic $(abc)^{\frac{1}{2}}S_0 + \theta S_1 = 0$, which passes through the common points of S_0 and S_1; and conversely. Further, any conic through the intersections of S_0 and S_1 has an equation of the form $\phi^2 S_0 - S_1 = 0$. We prove that this conic is the locus of the intersection of two tangents, respectively of the conics S_0 and S_1, say p and z, which are conjugate to one another in regard to the conic $\phi \Sigma_0 + (abc)^{\frac{1}{2}} \Sigma_1 = 0$, which touches the common tangents of S_0 and S_1. But further, when $\phi = \theta$, the two dually corresponding facts arise in the same figure, the lines p and z being the tangents at the points P, Z, of the conics S_0 and S_1.

Let S_0, S_1 and S be any three conics which have four points in common. Let the tangent line, p, at any point, P, of the conic S_0, meet S in two points of which C is one. Let the two tangents be drawn from C to the conic S_1, the point of contact of one of these being Z. Denote the line PZ by c. From the second intersection, D, with the conic S, of the line p, draw a tangent to S_1, touching this in X. Let ZX meet p in Q. Let the line joining P to the intersection of the tangents of S_1 at X and Z be q.

Let the equations of S_0, S_1 be, respectively $S_0 = 0$, $S_1 = 0$, where $S_0 = x^2 + y^2 + z^2$, $S_1 = ax^2 + by^2 + cz^2$, and the equation of S be $\theta^2 S_0 - S_1 = 0$; let the coordinates of the point P be (ξ, η, ζ), and of the line p be (λ, μ, ν), so that we have $\xi^2 + \eta^2 + \zeta^2 = 0$, $\lambda^2 + \mu^2 + \nu^2 = 0$, $\lambda\xi + \mu\eta + \nu\zeta = 0$. Let (x, y, z) be any other point, and (l, m, n) any other line; let $P_0 = x\xi + y\eta + z\zeta$, and $P_1 = ax\xi + by\eta + cz\zeta$. Also let $\Sigma_0 = l^2 + m^2 + n^2$, $\Sigma_1 = a^{-1}l^2 + b^{-1}m^2 + c^{-1}n^2$, $\Pi_0 = l\lambda + m\mu + n\nu$, $\Pi_1 = a^{-1}l\lambda + b^{-1}m\mu + c^{-1}n\nu$. We can at once verify the following equations, identically true in regard to x, y, z and l, m, n:

$$(\theta^2 - a)(by\zeta - cz\eta)^2 + (\theta^2 - b)(cz\xi - ax\zeta)^2 + (\theta^2 - c)(ax\eta - by\xi)^2$$
$$= -(bc\xi^2 + ca\eta^2 + ab\zeta^2)S_1 + abcP_0^2 - \theta^2 P_1^2,$$

$$abc\{(a^{-1}\theta^2 - 1)(y\zeta - z\eta)^2 + (b^{-1}\theta^2 - 1)(z\xi - x\zeta)^2$$
$$+ (c^{-1}\theta^2 - 1)(x\eta - y\xi)^2\}$$
$$= \theta^2 (a\xi^2 + b\eta^2 + c\zeta^2)S_1 + abcP_0^2 - \theta^2 P_1^2,$$

$$(\theta^{-2} - a^{-1})(b^{-1}m\nu - c^{-1}n\mu)^2 + (\theta^{-2} - b^{-1})(c^{-1}n\lambda - a^{-1}l\nu)^2$$
$$+ (\theta^{-2} - c^{-1})(a^{-1}l\mu - b^{-1}m\lambda)^2$$
$$= -[(bc)^{-1}\lambda^2 + (ca)^{-1}\mu^2 + (ab)^{-1}\nu^2]\Sigma_1 + (abc)^{-1}\Pi_0^2 - \theta^{-2}\Pi_1^2,$$

$$(abc)^{-1}\{(a\theta^{-2} - 1)(m\nu - n\mu)^2 + (b\theta^{-2} - 1)(n\lambda - l\nu)^2$$
$$+ (c\theta^{-2} - 1)(l\mu - m\lambda)^2\}$$
$$= \theta^{-2}(a^{-1}\lambda^2 + b^{-1}\mu^2 + c^{-1}\nu^2)\Sigma_1 + (abc)^{-1}\Pi_0^2 - \theta^{-2}\Pi_1^2,$$

which we shall call, respectively, (1), (2), (3) and (4). It will be seen that (3) and (4) are at once deducible from (1) and (2), and it is easy to deduce (2) from (1).

Now, regarding x, y, z as the coordinates of a variable point, the left side of the equation (1) vanishes when the point, which is the intersection of the line p with the polar line of (x, y, z) in regard to S_1, lies on the conic $\theta^2 S_0 - S_1 = 0$, or S; that is, when (x, y, z) is the pole, in regard to S_1, of any line through C or D. In other words, the left side of equation (1), put equal to zero, is the equation of the line pair consisting of the polars, in regard to S_1, of the points C and D. These lines pass, respectively, through Z and X, and meet S_1 in the points of contact of the other tangents to S_1 drawn from C and D, say Y and T. On the right side of equation (1), the terms $abc P_0^2 - \theta^2 P_1^2$, put equal to zero, would represent the line pair consisting of the polars of (ξ, η, ζ) in regard to the two conics $(abc)^{\frac{1}{2}} S_0 \pm \theta S_1 = 0$. Thus, the right side of equation (1), put equal to zero, would represent a conic passing through the points where these two polar lines meet the conic S_1. Therefore, these two polar lines are either the pair XZ, YT, or else the pair XY, ZT. We prove that they are the former pair. For we have shewn (Chap. I, Ex. 30) that the tangents of the conic S_1, at the points where it is met by the line PZ, are such as to meet the tangent, p, of S_0, in points lying on a conic through the common points of S and S_1; and such conic is identified by the single point C in addition to these four common points. Thus P, Z, T are in line, and, similarly, P, X, Y are in line. Wherefore, the three points consisting of the point P, the point of intersection, say, O, of XT and YZ, and the point of intersection of ZX and TY, form a self-polar triad in regard to S_1. Then, as O, the intersection of the polars of C and D in regard to S_1, is the pole of CD, it follows that ZX and TY meet on CD, so that Q, Y, T are in line. The lines whose equations are $(abc)^{\frac{1}{2}} P_0 \pm \theta P_1 = 0$, not containing the point P, save for one particular value of θ, are thus the lines ZX, TY. Therefore, the point Q is the point of intersection of the polars of P taken in regard to all the conics passing through the common points of S_0 and S_1. It follows, thence, that the points P, X are harmonic conjugates of one another in regard to the two points in which the line PX meets one of the two conics $(abc)^{\frac{1}{2}} S_0 \pm \theta S_1 = 0$, while the points P, Y are harmonic conjugates of one another in regard to the two points in which this line meets the other of these two conics. If P' be the other point in which the line PX meets the conic S_0, it will appear that P' and Y are conjugate in regard to the same conic as that for which P, X are conjugate, while P' and X and P, Y are, likewise, two pairs of conjugate points with respect to the same one of the two conics $(abc)^{\frac{1}{2}} S_0 \pm \theta S_1 = 0$.

Next, consider the equation (2). The left side vanishes when the line joining the points (x, y, z) and (ξ, η, ζ) touches the conic whose tangential equation is $\theta^2 \Sigma_1 - \Sigma_0 = 0$. This is a particular conic touching the common tangent lines of S_0 and S_1. The left side of equation (2), put equal to zero, thus represents the two tangents drawn from P to the conic $\theta^2 \Sigma_1 - \Sigma_0 = 0$. The right side of equation (2), put equal to zero, represents, however, a conic passing through the points X, Y, Z, T; this consists, then, of the lines PXY, PZT. These lines, therefore, touch the conic $\theta^2 \Sigma_1 - \Sigma_0 = 0$, which we may denote by Σ; we may denote the lines PXY, PZT, respectively, by d and c.

Now consider the equation (3). Regarding l, m, n as coordinates of a variable line, the left side vanishes when the pole $(a^{-1}l, b^{-1}m, c^{-1}n)$, in regard to Σ_1, of this line, and the pole, P, or (λ, μ, ν), of the line p, determine a line which touches the conic $\theta^{-2} \Sigma_0 - \Sigma_1 = 0$, which is the conic Σ. In other words, the left side of equation (3) vanishes for all lines (l, m, n) which are polars, in regard to Σ_1, of points of either of the two lines PXY, PZT; thus, put equal to zero, it represents, tangentially, the point pair constituted by the poles, in regard to Σ_1, of the lines d and c; if the tangents of S, at the points X, Y, Z, T, be called, respectively, x, y, z, t, these poles are the points (x, y) and (z, t). On the right side of the equation (3), the terms $(abc)^{-1} \Pi_0^2 - \theta^{-2} \Pi_1^2$ vanish when the line (l, m, n) passes through the pole, in regard to one of the conics $(abc)^{-\frac{1}{2}} \Sigma_0 \pm \theta^{-1} \Sigma_1 = 0$, of the line (λ, μ, ν); that is, these terms, put equal to zero, are the tangential equation of the point pair consisting of the poles of the line p in regard to these two conics. The right side of equation (3), put equal to zero, is, tangentially, the equation of a conic touching the tangents drawn from these poles to the conic S_1. We thus infer that the poles of the line p, in regard to the conics $\theta \Sigma_0 \pm (abc)^{\frac{1}{2}} \Sigma_1 = 0$, are a complementary pair of intersections of the tangents drawn from C and D to the conic S_1, other than the pair (x, y) and (z, t); they are also other than C and D. These poles are then the points (x, z) and (y, t). It is known that these points lie on a line, q, passing through P. Thus, the tangent, z, drawn from C to the conic S_1, is the conjugate of the tangent, p, drawn from C to the conic S_0, in regard to one of the conics $\theta \Sigma_0 \pm (abc)^{\frac{1}{2}} \Sigma_1 = 0$, and the other tangent, y, is the conjugate of p in regard to the other of these conics. It will appear that the other tangent drawn from C to the conic S_0 may be similarly coupled, respectively, with y and z.

Lastly, consider the equation (4). The left side, equated to zero, is the tangential equation of lines meeting the line p upon the conic $S_1 - \theta^2 S_0 = 0$, or the conic S. Thus it represents the point pair C, D. The equation thus expresses the fact we have already found, that the tangents from the points (x, z), (y, t), to the conic S_1, have the points, C, D, as a complementary pair of intersections.

We have thus proved the results stated at the beginning of the example. We consider some particular cases.

A particular case arises when the conic $(abc)^{\frac{1}{2}} S_0 - \theta S_1 = 0$ is a pair of common chords of the conics S_0 and S_1, so that, in brief, the line PZ is divided harmonically by these two lines. The common chords will intersect in one of the three points which form the common self-polar triad for these two conics S_0, S_1. The line joining the other two points of this triad will contain two complementary intersections each of two of the common tangents of S_0 and S_1. It can be shewn that, in this case, the envelope $\Sigma_0 - \theta^2 \Sigma_1 = 0$ touches both the common chords considered, and that the conic $\theta^2 S_0 - S_1 = 0$ passes through the two intersections of common tangents spoken of. For, when $(abc)^{\frac{1}{2}} S_0 - \theta S_1 = 0$ is a line pair, say, for $\theta = (bc/a)^{\frac{1}{2}}$, the envelope $\theta \Sigma_0 - (abc)^{\frac{1}{2}} \Sigma_1 = 0$ is the point pair consisting of the two intersections of common tangents spoken of. The coordinates (l, m, n) of the line pair $(abc)^{\frac{1}{2}} S_0 - (bc/a)^{\frac{1}{2}} S_1 = 0$, are, respectively, $l = 0$, $m = (c - a)^{\frac{1}{2}}$, $n = \pm (a - b)^{\frac{1}{2}}$; these touch $\Sigma_0 - (bc/a)\Sigma_1 = 0$, which is $a^{-1}(a^2 - bc) l^2 - (c - a) m^2 + (a - b) n^2 = 0$. The common tangents of S_0 and S_1 have coordinates $[a(b - c)]^{\frac{1}{2}}$, $\pm [b(c - a)]^{\frac{1}{2}}$, $[c(a - b)]^{\frac{1}{2}}$; when $\theta = (bc/a)^{\frac{1}{2}}$, the envelope $\theta \Sigma_0 - (abc)^{\frac{1}{2}} \Sigma_1 = 0$ reduces to $[b(c - a)]^{-1} m^2 - [c(a - b)]^{-1} n^2 = 0$, which represents the two points of intersection of common tangents spoken of. The coordinates of these two points are $x = 0$, $y = [b(c - a)]^{-\frac{1}{2}}$, $z = \pm [c(a - b)]^{-\frac{1}{2}}$; these points lie on the conic $\theta^2 S_0 - S_1 = 0$, which is

$$(bc - a^2) x^2 + b(c - a) y^2 - c(a - b) z^2 = 0.$$

If the two intersections of common tangents spoken of, be called U and V, it appears that the tangents CP, CZ, respectively to S_0 and S_1, drawn from C, are harmonic conjugates in regard to CU and CV; if these tangents meet the conic $\theta^2 S_0 - S_1 = 0$ in further points D and Z', these points will, then, be harmonic conjugates in regard to the points U, V, of this conic. The line DZ' will, thus, pass through the pole of the chord UV of this conic; this is that one originally considered of the points, which are the common self-polar triad of S_0 and S_1, through which the common chords of this conic pass. Thus we have triads of points, C, D, Z', lying on the conic S, of which two joins touch, respectively, the conics S_0 and S_1, while the remaining join, DZ', passes through one of the triad of common conjugate points for S_0 and S_1. (In the general case, Chap. I, Ex. 31, the line DZ' touches a certain conic through the common points of S_0 and S_1.) As, however, the particular case in which the conic $(abc)^{\frac{1}{2}} S_0 - \theta S_1 = 0$ is a pair of common chords of S_0 and S_1, is, as has been remarked, also the particular case in which the envelope $\theta \Sigma_0 - (abc)^{\frac{1}{2}} \Sigma_1 = 0$ is a pair of intersections of common tangents of S_0 and S_1, the fact that $\Sigma_0 - \theta^2 \Sigma_1 = 0$ touches the common chords corresponds dually to the fact that the conic

$\theta^2 S_0 - S_1 = 0$ passes through these intersections of common tangents. We have remarked that the tangents CP, CZ are harmonic in regard to CU and CV; these last are, in fact, the double rays of a pencil in involution, of which CP, CZ are a pair, and two other pairs are the lines drawn from C to the pairs of intersections of S_0 and S_1, two intersections being paired when they lie on one of the common chords which we considered. This is obvious because all these four intersections lie on the conic $\theta^2 S_0 - S_1 = 0$ which contains C, and the chords which join these pairs pass through the pole, in regard to this conic, of the line UV. Dually, if we denote the common chords considered by u and v, and consider the envelope $\Sigma_0 - \theta^2 \Sigma_1 = 0$ for the particular value of θ necessary that the line c may be a tangent, the tangents, u, v, of this envelope, meet the tangent c in

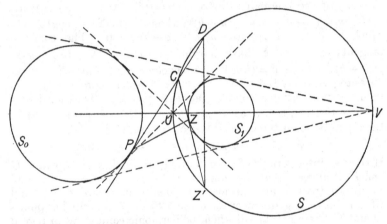

the double points of an involution, of which one pair are the points where c meets S_0 and S_1, and another pair are the points where c meets either pair of common tangents of S_0 and S_1 which meet on the line UV.

We may suppose the conics S_0, S_1 to be two circles, the common chords considered being the radical axis, and the line IJ. Then, if U, V be the intersections of the common tangents of the two circles which lie on the line of centres, the circle, S, coaxial with S_0 and S_1, which passes through U and V, is such that tangents CP, CZ, drawn from a point, C, of S, respectively to S_0 and S_1, meet S again in points, D, Z', lying on a line at right angles to the line of centres; further, the lines joining C to the absolute points I, J, common to the circles, as, also, the lines joining C to the other intersections of the circles, are harmonic in regard to CU and CV. Also, the line PZ is bisected by the radical axis; and this line envelopes a parabola which touches the radical axis, and touches the common tangents of the circles, and has for focus the middle point of the centres of the two circles. The two common tangents which meet in U are met by PZ in two points whose middle point is on the radical axis, as are the two common tangents which meet in V.

We may also regard the conics Σ_0, Σ_1 as two conics of a confocal system, the points of intersection, U, V, of pairs of common tangents, being taken for the absolute points, I, J. There is, then, a circle, passing through the common points of the two confocals, at every point of which intersect a tangent of one conic, and a tangent of the other at right angles to this. The two common chords of the confocals which meet in their centre subtend right angles at any point of this circle. The line joining the points of contact of the

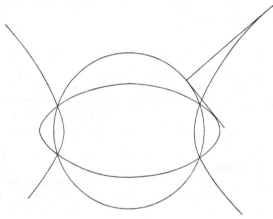

two perpendicular tangents envelopes a conic confocal with the given ones, which touches the common central chords of these ; this line is divided harmonically by these chords, and divided harmonically by the lines joining I (or J) to the two foci, S, H, of the two conics.

A still more particular case of the general theorem is the fact that if a line be drawn, from a focus, S, of a conic, at right angles to the tangent, at a point, P, of the conic, meeting this in N, then N describes a circle, whose centre is the centre of the conic, having two points of contact with the conic. If the point equation of the conic be $a^{-1}x^2 + b^{-1}y^2 - z^2 = 0$, the lines NP, NS are tangents of the two envelopes expressed by $al^2 + bm^2 - n^2 = 0$ and $(a - b)\,l^2 - n^2 = 0$, and are harmonic conjugates in regard to the two lines drawn from N which belong to the envelope expressed by

$$al^2 + bm^2 - n^2 - [(a - b)\,l^2 - n^2] = 0, \text{ or by } l^2 + m^2 = 0 ;$$

this envelope is the point pair I, J. The point N thus lies on the locus whose point equation is $a^{-1}x^2 + b^{-1}y^2 - z^2 - (b^{-1} - a^{-1})y^2 = 0$, or $x^2 + y^2 - az^2 = 0$; this is the auxiliary circle of the conic. (See Chap. II, p. 86, above.)

Ex. 19. Indication of a generalisation of the preceding results. Referring back to the result given above in Ex. 4 (p. 111), we see that, if from the four elements defined by the two roots of the quadratic $f = 0$, and the two roots of the quadratic $f' = 0$, we form two pairs, by taking each root of $f = 0$ with one of the roots of $f' = 0$, which is possible in two ways, then the elements which are harmonic in regard to the constituents of both the pairs are given by one of the equations $(\Delta')^{\frac{1}{2}}f \pm (\Delta)^{\frac{1}{2}}f' = 0$. But the result of Ex. 4 leads to a result for two conics of which the corresponding corollary is in connexion with the result of the preceding Example 18. Let $S = 0$, $\Sigma = 0$ be the point equation, and the tangential equation, of the same conic, where

$$S = ax^2 + \ldots + 2fyz + \ldots,$$
$$\Sigma = -(bc - f^2)u^2 - \ldots - 2(gh - af)vw - \ldots,$$

let S', Σ' be what these become when a', b', ... are written, respectively, for a, b, ...; also let $T = 0$ be the condition for a line, of coordinates u, v, w, whose intersections with $S = 0$ are harmonic in regard to its intersections with $S' = 0$ (above, Ex. 17), where

$$T = -(bc' + b'c - 2ff')u^2 - 2(gh' + g'h - af' - a'f)vw - \ldots.$$

Then, let the line (u, v, w) meet $S = 0$ in the points A, B, meet $S' = 0$ in the points A', B', and meet $\theta S + \theta'S' = 0$ in the points X, Y. In terms of the symbols of A, A', let the symbols of X and Y be $X = A + \xi A'$ and $Y = A + \eta A'$; and let $\epsilon = \xi^{-1}\eta$. It follows from Ex. 4 that

$$\frac{4\epsilon}{(1 + \epsilon)^2} = \frac{\theta\theta'[2(\Sigma\Sigma')^{\frac{1}{2}} - T]}{[\theta(\Sigma)^{\frac{1}{2}} + \theta'(\Sigma')^{\frac{1}{2}}]^2}.$$

Thus, in particular, when X, Y are harmonic in regard to A, A', and so $\epsilon = -1$, the line (u, v, w) envelopes the conic whose tangential equation is $\theta^2\Sigma - \theta'^2\Sigma' = 0$.

Ex. 20. Of correspondences, in particular of general involutions, upon a conic. Taking the conic of which any point has coordinates of the form θ^2, θ, 1, we may consider two points, P and Q, of the conic, with parameters, respectively, θ and ϕ, which are so related that there exists an equation of the form

$$\theta^m(a\phi^n + b\phi^{n-1} + \ldots) + \theta^{m-1}(a_1\phi^n + b_1\phi^{n-1} + \ldots) + \ldots$$
$$+ a_m\phi^n + b_m\phi^{n-1} + \ldots = 0,$$

wherein the polynomial on the left is supposed to be general, containing every term $\theta^r\phi^s$ with $0 \gtreqless r \gtreqless m$ and $0 \gtreqless s \gtreqless n$. A particular case is that in which m and n are both 1, so that P and Q describe related ranges upon the conic (Vol. I, p. 140). In general, we say that there is a (m, n) correspondence between P and Q, there being m positions of P corresponding to any position of Q, and n positions of Q corresponding to any position of P. In saying this we are again assuming that a polynomial expression in a parameter vanishes for a number of values of this parameter equal to the degree of the polynomial. See Chap. IV, below.

As a rule, it is not the case, in such a correspondence, that when Q takes the position P, then P takes one of the corresponding positions Q. This will be so if $m = n$, and the equation connecting θ, ϕ be symmetrical in regard to these. When $m = n = 1$ this arises if P, Q are pairs of an involution upon the conic. When $m = n = 2$ it arises if the relation be of the form

$$\theta^2 (a\phi^2 + h\phi + g) + \theta (h\phi^2 + b\phi + f) + g\phi^2 + f\phi + c = 0 ;$$

then the two positions of P which correspond to any position of Q are the intersections of the fundamental conic, $xz - y^2 = 0$, with the polar of Q in regard to the conic $ax^2 + 2fyz + \ldots = 0$. The polar of P in regard to this last conic then passes through Q, but meets the fundamental conic, in general, in a new point, R; thus, when Q takes the position P, one of the corresponding positions of P coincides with Q but the other is generally a fresh point. Another interpretation of the relation is obtained by remarking that, if the line joining P and Q, be written $ux + vy + wz = 0$, so that $u = 1$, $v = -(\theta + \phi)$, $w = \theta\phi$, we have

$$aw^2 - hvw + g(v^2 - 2wu) + bwu - fuv + cu^2 = 0,$$

and the line PQ envelopes a conic, of which this is the tangential equation.

A particular case of the general symmetrical correspondence just described, is that in which any point Q, and the m points P which correspond to it, form a set of $(m + 1)$ points which remains the same set when Q takes the position of any one of the $(m + 1)$ points which form the set. Such a series of sets of $(m + 1)$ points is called an involution of sets of $(m + 1)$ points, and is of great importance. In the case $m = 2$, $n = 2$, this, clearly, requires that it be possible to take triads of points of $xz - y^2 = 0$ which are self-polar in regard to $ax^2 + 2fyz + \ldots = 0$, or that it be possible to take triads of points of $xz - y^2 = 0$ whose joins touch the conic whose tangential equation is that written above. We have previously seen (Chap. i, Exx. 12 and 7) that, in either case, if one such triad be possible, an infinite number of such triads is possible. For general values of m, it is convenient to consider the equation satisfied by all the $(m + 1)$ points forming the symmetrical set consisting of Q and the m corresponding points P. It can be shewn that this equation must necessarily be of the form

$$\theta^{m+1} (a + \lambda b) + \theta^m (a_1 + \lambda b_1) + \ldots + \theta (a_m + \lambda b_m) + a_{m+1} + \lambda b_{m+1} = 0,$$

where a, b, a_1, b_1, ... are constants, but λ varies from set to set. Thus the points of a set are given by values of θ satisfying an equation $u + \lambda v = 0$, but the points of another set by values of θ satisfying an equation $u + \lambda' v = 0$; here u, v are two definite polynomials in θ of degree $m + 1$, which, corresponding to $\lambda = 0$, $\lambda^{-1} = 0$, respectively, give the points of two particular sets of the series. The whole involution is thus determined by two of its sets.

We may give the proof of this result : let the m points P corresponding to Q be given by the equation

$$\theta^m u_0 + \theta^{m-1} u_1 + \ldots + u_m = 0,$$

where u_0, u_1, ..., u_m are polynomials in the parameter, ϕ, by which Q is determined. By the definition of the involution, if we replace θ, in this equation, by one of its roots, say θ_1, and ϕ by any other of the roots, say θ_2, the equation remains true. Thus, forming the equation

$$(\sigma - \phi) (\sigma^m u_0 + \sigma^{m-1} u_1 + \ldots + u_m) = 0,$$

or, say,

$$\sigma^{m+1} u_0 + \sigma^m v_1 + \sigma^{m-2} v_2 + \ldots + v_{m+1} = 0,$$

each of u_0, v_1, v_2, ..., v_{m+1} is a polynomial in ϕ, and this equation is satisfied not only by $\sigma = \phi$, but, also, by each of the values, $\sigma = \theta_1$, $\sigma = \theta_2$, ..., $\sigma = \theta_m$, which, with ϕ, make up the set of $m + 1$ values belonging to a set. These $(m + 1)$ roots are, however, also obtained from the equation

$$\sigma^{m+1} u_0' + \sigma^m v_1' + \sigma^{m-2} v_2' + \ldots + v'_{m+1} = 0$$

which differs from the preceding in having any one of θ_1, θ_2, ..., θ_m written in place of ϕ, in the coefficients; for instance, If then we take a value of r for which v_r/u_0 does actually contain ϕ, and denote this by $F(\phi)$, we have $F(\phi) = F(\theta_1) = F(\theta_2) = \ldots = F(\theta_m)$. Thus the $m + 1$ points of any set are given by $F(\phi) = \lambda$, for a proper value of λ; it follows, since u_0 is at most of order m in ϕ, that v_r, in $F(\phi) = v_r/u_0$, must be of order $m + 1$, at least, if the values ϕ, θ_1, θ_2, ..., θ_m are all different; and it cannot be of higher order, since none of u_1, u_2, ..., u_m is of higher than order m in ϕ. We shall assume that ϕ, θ_1, ..., θ_m are all different, save for particular values of ϕ.

If s be another suffix such that v_s/u_0 does actually contain ϕ, so that, by the same argument, v_s is of order $m + 1$ in ϕ, and the values $\phi, \theta_1, \ldots, \theta_m$ are given by the roots of $v_s/u_0 = \mu$, for a proper value of μ, it can be deduced that $v_s/u_0 = AF(\phi) + B$, where A, B do not depend upon ϕ. The equation $\sigma^{m+1}u_0 + \sigma^m v_1 + \ldots = 0$ can then be arranged in the form $P + \lambda Q = 0$, where P, Q are polynomials in σ, not containing ϕ, and λ is put for $F(\phi)$.

We may apply the preceding theory to find when two points $(\theta^2, \theta, 1)$, $(\phi^2, \phi, 1)$, of the conic $xz - y^2 = 0$, which are in $(2, 2)$ correspondence expressed by the equation before written,

$$\theta^2(a\phi^2 + h\phi + g) + \ldots = 0,$$

or, say $F(\theta, \phi) = 0$, are two points of a set of three points belonging to a series in involution. For this it is necessary, and sufficient, that there should be two polynomials of the third degree, in the parameter x, say $u(x)$, $w(x)$, such that the $(2, 2)$ relation above written, $F(\theta, \phi) = 0$, is obtainable by elimination of λ from the two equations $u(\theta) + \lambda w(\theta) = 0$, $u(\phi) + \lambda w(\phi) = 0$, the equation $F(\theta, \phi) = 0$ being then of the form $(\theta - \phi)^{-1}[u(\theta) w(\phi) - u(\phi) w(\theta)] = 0$. But, if the two values of θ, corresponding to ϕ in virtue of $F(\theta, \phi) = 0$, taken with ϕ itself, constitute such a set, the equation in σ, $(\sigma - \phi) F(\sigma, \phi) = 0$, will give the set; therefore, by the theory given above, the various sets can be found from the equation $v_3/u_0 = \lambda$, namely

$$-\phi(g\phi^2 + f\phi + c)/(a\phi^2 + h\phi + g) = \lambda,$$

for varying values of λ. In particular, for $\lambda^{-1} = 0$, one set will be that given by $\phi^{-1} = 0$, and $a\phi^2 + h\phi + g = 0$, and another set will be that given by $\phi(g\phi^2 + f\phi + c) = 0$. If we write

$$u(\phi) = a\phi^2 + h\phi + g, \quad w(\phi) = \phi(g\phi^2 + f\phi + c),$$

it is easy to verify the identity

$$(\phi - \theta)^{-1}[u(\theta) w(\phi) - u(\phi) w(\theta)]$$
$$= gF(\theta, \phi) + [hf - bg - (ca - g^2)]\phi\theta.$$

Therefore, we see, supposing g not to be zero, that the condition that the $(2, 2)$ relation in question should connect two points of a set of three points of such a series in involution, is

$$hf - bg = ca - g^2.$$

This condition can also be shewn, if we write $(\sigma - \phi) F(\sigma, \phi)$ in the form $u(\phi) . (\sigma^3 + H\sigma^2 + K\sigma + \lambda)$, to secure that $H = g^{-1}(a\lambda + f)$, $K = g^{-1}(h\lambda + c)$, in accordance with the theory given above. It will be obtained below (Ex. 24) as the condition that it be possible to find triads of points of the conic $xz - y^2 = 0$ which are self-polar in regard to the conic $ax^2 + 2fyz + \ldots = 0$.

Taking a fixed point, O, of the conic $xz - y^2 = 0$, and three other fixed points, A, B, C, not lying on this conic, the general conic through these four points will have an equation of the form $U + \lambda V = 0$. The three intersections of this conic with $xz - y^2 = 0$, other than O, will then be given by an equation $u + \lambda v = 0$, where u, v are cubic polynomials in the parameter, ϕ, of a point of $xz - y^2 = 0$; these three intersections will then be a set of an involution lying on this conic, of which the different sets are given by different values of λ. Conversely, let P, Q, R and P', Q', R' be any two triads of points of this conic; take, upon the conic, an arbitrary fixed point, O; then draw any definite conic through P, Q, R, O, and any other definite conic through P', Q', R', O; these two conics will have in common three further points, A, B, C, not generally lying upon the given conic $xz - y^2 = 0$. The general conic through A, B, C, O will then meet this given conic in three further points P'', Q'', R'', other than O, and these will belong to the involution of sets of three points determined by the two sets P, Q, R and P', Q', R'. Evidently, in this construction, the point O, and one of the points A, B, C may be taken arbitrarily; in particular, denoting

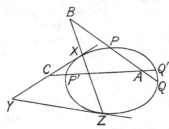

by X, Z the points of contact of the given conic, $xz - y^2 = 0$, with $z = 0$, $x = 0$, respectively, there will be a set of the involution containing X; say this consists of X, P, Q; and there will be a set, say Z, P', Q', containing Z; assuming that Z is not one of P, Q and X is not one of P', Q', we may write the equation of the lines $PQ, P'Q'$ in the respective forms $ax + hy + gz = 0$, $gx + fy + cz = 0$. Then the sets of the involution will be given by

$$\lambda \,(a\phi^2 + h\phi + g) + \phi \,(g\phi^2 + f\phi + c) = 0,$$

for varying values of λ, the points X, Z being given by $\phi^{-1} = 0$, $\phi = 0$, respectively; this is the same as given above. But this involution is obtained by the intersections of the conic $xz - y^2 = 0$ with the conics of equation $\lambda z\,(ax + hy + gz) + y\,(gx + fy + cz) = 0$, for varying λ. These conics have for four common points; first, the point X, on $xz - y^2 = 0$; then the point, A, where the lines, PQ, $P'Q'$ meet; then B, where XZ, PQ meet; and, finally, C, where XY and $P'Q'$ meet.

A particular result, referred to in Chap. II (above, p. 87), furnishes an example of the theorem obtained here, that the sets of an involution, of three points, are given by an equation $\lambda u + v = 0$, for varying values of λ, where u, v are cubic polynomials in the parameter which determines a point. For, if we assume, with anticipation of a limitation introduced in the following Chap. IV, that a cubic equation, in σ, whose roots are a, b, c, is capable of the form $(\sigma - a)(\sigma - b)(\sigma - c) = 0$, this shews that if a, b, c be the values belonging to any set of the involution, we have equations $a + b + c = P + Q abc$, $bc + ca + ab = R + S abc$, where P, Q, R, S are the same for all the sets.

Now consider the pedal lines of a fixed point of the circle $x^2 + y^2 = z^2$ (the absolute points being $z = 0$, $x \pm iy = 0$), taken in regard to the joins of the three points of the circle which are a set of an involution of sets of three points lying on the circle. We shew that, for all these sets, the pedal lines have a point in common. Putting $x + iy = X$, $x - iy = Y$, so that the circle is $XY = z^2$, the parameter for a point of the circle may be taken to be $\theta = X/z$. If for the fixed point this be λ, and for the points of a set of the involution the values be $\theta = a$, $\theta = b$, $\theta = c$, it is easy to prove that the equation of the pedal line is

$$2\lambda(Y abc - X\lambda) = [abc + \lambda(bc + ca + ab) - \lambda^2(a + b + c) - \lambda^3]z;$$

with the conditions given above, the right side is, save for the factor z,

$$abc(1 + \lambda S - \lambda^2 Q) + \lambda R - \lambda^2 P - \lambda^3;$$

thus the pedal line passes through the fixed point given by

$$Xz^{-1} = \tfrac{1}{2}(\lambda + P - \lambda^{-1}R), \quad Yz^{-1} = \tfrac{1}{2}(-\lambda Q + S + \lambda^{-1}).$$

(Third year problems, St John's College, Cambridge, Dec. 1889. Cf. Weill, *Liouville's J.* IV, 1878, 270.)

In general, in an involution of sets of n points, there will be $(n - 1)^2$ sets whereof two of the n points coincide, as is easy to see. For $n = 3$, taking two such sets as the fundamental sets, the parameter being chosen so that the points of coincidence are given by $\theta = 0$, $\theta = \infty$, the involution may be represented by

$$\theta^2(b\theta + a) + \lambda(c\theta + b) = 0.$$

Now suppose the points of the involution to be points, $(\theta^2, \theta, 1)$, of the conic $xz - y^2 = 0$. Define the pole of a line in regard to points A, B, C as the point O such that AO and the line meet BC in points harmonic in regard to B and C, with a similar condition for C, A and for A, B. Then we may prove that, for all sets A, B, C of the involution $\theta^2(b\theta + a) + \lambda(c\theta + b) = 0$, the pole of the line joining the two double points, $\theta = 0$, $\theta = \infty$, is the same, being the point $(a, -3b, c)$.

The condition for an involution of sets of three points upon a conic may be approached differently. The following examples lead up to this.

Ex. 21. *Reciprocation of one conic into another when they have a common self-polar triad.* As has already been remarked (above,

p. 107), the equations of two conics which have a common self-polar triad, when the coordinates are taken in regard thereto, are of the forms

$$ax^2 + by^2 + cz^2 = 0, \quad a'x^2 + b'y^2 + c'z^2 = 0.$$

Now let $Ax^2 + By^2 + Cz^2 = 0$ be a further conic for which the same triad is self-polar. If we consider the locus of the poles, taken in regard to this last conic, of the tangents of one of the former conic, it will be another conic. In fact the pole of the line $ax\xi + by\eta + cz\zeta = 0$, in regard to $Ax^2 + \ldots = 0$, is $(A^{-1}a\xi, B^{-1}b\eta, C^{-1}c\zeta)$, and when (ξ, η, ζ) is such that $a\xi^2 + b\eta^2 + c\zeta^2 = 0$, this pole describes the conic

$$a^{-1}A^2x^2 + b^{-1}B^2y^2 + c^{-1}C^2z^2 = 0.$$

If we consider the envelope of the polars, also taken in regard to $Ax^2 + \ldots = 0$, of the points of the same conic, $ax^2 + \ldots = 0$, it is similarly seen to consist of the tangents of the conic $a^{-1}A^2x^2 + \ldots = 0$. This last is then called the polar reciprocal of $ax^2 + \ldots = 0$ in regard to $Ax^2 + \ldots = 0$. There are thus four conics $Ax^2 + \ldots = 0$ in regard to which the polar reciprocal of $ax^2 + \ldots = 0$ is the other given conic $a'x^2 + \ldots = 0$, namely those for which $A^2 = aa'$, $B^2 = bb'$, $C^2 = cc'$.

Conversely, if a conic S be reciprocated to a conic S', a self-polar triad of points for S becomes a self-polar triad of lines for S', as we easily see; thus, when S and S' have a common self-polar triad, this is also self-polar for the reciprocating conic.

It may be remarked that if we take the polar line of a point, P, or (ξ, η, ζ), in regard to the conic $Ax^2 + \ldots = 0$, so obtaining $Ax\xi + By\eta + Cz\zeta = 0$, and then take the pole of this line in regard to one of the three associated conics, say $-Ax^2 + By^2 + Cz^2 = 0$, we obtain the point $(-\xi, \eta, \zeta)$ or P', which can be obtained from P by joining the point $(1, 0, 0)$, or A, to P, by a line, AP, meeting BC in D, and taking P' on AD harmonic to P in regard to A and D.

Ex. 22. The determination of the common self-polar triad of two conics in general. The construction for the common self-polar triad of two conics has been given (above, p. 23) when their four common points are given, and these are distinct. When the equations of the conics are given, in general form, say $ax^2 + 2fyz + \ldots = 0$, $a'x^2 + 2f'yz + \ldots = 0$, the algebraic determination will begin by seeking a point, (x, y, z), whose polar line is the same for both conics. For this, we must have

$$\frac{ax + hy + gz}{a'x + h'y + g'z} = \frac{hx + by + fz}{h'x + b'y + f'z} = \frac{gx + fy + cz}{g'x + f'y + c'z};$$

these lead to three equations of the form

$$(a - \lambda a')x + (h - \lambda h')y + (g - \lambda g')z = 0,$$

which, if the two given conics be $S = 0$, $S' = 0$, express that the polar lines of the three points $(1, 0, 0)$, etc., taken in regard to the conic $S - \lambda S' = 0$, meet in a point. Thus the conic $S - \lambda S' = 0$, for the appropriate value of λ, is a pair of lines, through the common points of the given conics, which intersect in (x, y, z). By elimination of x, y, z we find that λ is a root of a cubic equation, say $\Delta(\lambda) = 0$, which arises naturally in determinant form. When the given conics have four distinct intersections, there are three line pairs through these, and, therefore, this cubic equation has three roots which are different. Denote them by $\lambda_1, \lambda_2, \lambda_3$, in this case. Then let l_1, m_1, n_1 be determined from two of the equations of the form

$$(a - \lambda_1 a')\, l_1 + (h - \lambda_1 h')\, m_1 + (g - \lambda_1 g')\, n_1 = 0,$$

the third of these equations being a consequence of these two, and, similarly, let l_2, m_2, n_2 be found from the equations of the same form in which λ_1 is replaced by λ_2 and l_1, m_1, n_1 by l_2, m_2, n_2, respectively. If we denote by S_{12} the expression $al_1 l_2 + f(m_1 n_2 + m_2 n_1) + \ldots$, a sum of six terms, which is symmetrical in regard to (l_1, m_1, n_1) and (l_2, m_2, n_2), and denote by S'_{12} the similar expression in which $a, f, \ldots$ are replaced by $a', f', \ldots$, the first set of three equations enables us to deduce $S_{12} - \lambda_1 S'_{12} = 0$, and the second set of three equations similarly enables us to deduce $S_{12} - \lambda_2 S'_{12} = 0$. As we are supposing λ_1 and λ_2 to be different we can infer that $S_{12} = 0$, $S'_{12} = 0$. In the same way we infer that $S_{23} = S'_{23} = S_{31} = S'_{31} = 0$. From these it follows that if, in S and S', we replace x, y, z, respectively, by $l_1 \xi + l_2 \eta + l_3 \zeta$, $m_1 \xi + m_2 \eta + m_3 \zeta$, $n_1 \xi + n_2 \eta + n_3 \zeta$, then S has a form $P\xi^2 + Q\eta^2 + R\zeta^2$ and S' a form $P'\xi^2 + Q'\eta^2 + R'\zeta^2$; here, for instance, P is what S becomes when l_1, m_1, n_1 are written for x, y, z, with similar values for Q, R, P', Q', R', and, as is easy to see, $P = \lambda_1 P'$, $Q = \lambda_2 Q'$, $R = \lambda_3 R'$. As the equations from which l_1, m_1, n_1 are found are homogeneous in these, we may suppose them taken so that $P' = 1$, and, similarly, may suppose l_2, m_2, n_2 and l_3, m_3, n_3 so taken that $Q' = 1$ and $R' = 1$.

When the roots $\lambda_1, \lambda_2, \lambda_3$ of the cubic equation above are not all different, the previous theory, and, in particular, the equations such as $S_{12} = 0$, may cease to hold, and it may not be possible to choose ξ, η, ζ so that S and S' involve only squares of these. Without entering into the theory at present, which finds its place in a subsequent volume, we give below the forms, for all cases, to which two conics can be reduced (Ex. 25).

Ex. 23. Fundamental invariants for two conics. As the cubic equation for λ, which arises above, has the geometrical interpretation in regard to the line pairs which pass through the common points of the conics $S = 0$, $S' = 0$, the four coefficients in this equation have values which, save for a common factor, are independent of the base points in terms of which the coordinates are taken. If we denote the cubic, written at length, by $\Delta - \lambda\Theta + \lambda^2\Theta' - \lambda^3\Delta' = 0$, the expression Δ is that polynomial, homogeneously of the third degree in the coefficients of the conic S, whose vanishing is the condition for $S = 0$ to be a line pair (above, p. 102), while Δ' is the similar polynomial for S', the expression Θ is only of the second degree in the coefficients of S, but is of the first degree in the coefficients of S', being, explicitly, the sum of three pairs of terms such as $Aa' + 2Ff'$, where $A = bc - f^2$, $F = gh - af$, etc., which are the coefficients in the tangential equation of S, while Θ' is similarly related to S'. We proceed to find the geometrical significance of the equation $\Theta = 0$, or of $\Theta' = 0$.

Ex. 24. The algebraical condition for one conic to be outpolar to another, or to be triangularly circumscribed to another. It was proved in Chap. I (Ex. 11), and will appear independently here, that if two conics S and Σ' be such that a triad of points of S be self-polar in regard to Σ', then there is an infinite number of other triads of points of S with the same property; the polar, in regard to Σ', of any point, P, of S, meets S in points, Q, R, such that P, Q, R is self-polar in regard to Σ'. And, when this is so, there are triads of tangents of Σ' which are self-polar in regard to S, also in infinite number. It was suggested that S should be spoken of as outpolar to Σ', and Σ' as inpolar to S. The property is one relating to the points of the outpolar conic, S, and to the tangents of the inpolar conic, Σ'; and its algebraic condition involves, most conveniently, the coefficients in the point equation of the outpolar conic, and the coefficients in the tangential equation of the inpolar conic. The necessary and sufficient condition is linear in both these sets of coefficients, being, in fact, the vanishing of the sum of the three pairs of terms such as $aA' + 2fF'$, which, above, we have called Θ'. As the expression is of the first degree in the coefficients of S, it follows that, if $S_1, S_2, \ldots$ be conics all outpolar to Σ, so also is any conic $A_1S_1 + A_2S_2 + \ldots = 0$.

It was also proved in Chap. I (Ex. 7), and will appear here, that if two conics S and Σ' be such that a triad of points A, B, C exists on the former, whose joins, BC, CA, AB, touch the latter, then there is an infinite number of other such triads, the tangents drawn to Σ', from any point of S, meeting S again in two points whose join is a tangent of Σ'. The condition for two conics to be in this relation is again primarily one for the points of S, and the tangents of Σ'. This condition is in fact

$$\tfrac{1}{4}(aA' + 2fF' + \ldots)^2$$
$$= (bc - f^2)(B'C' - F'^2) + 2(gh - af)(G'H' - A'F') + \ldots,$$

where, on both sides of the equation, there is a sum of three such pairs of terms as are written. If we suppose that the point equation, and not the tangential equation, of Σ', is given, being $a'x^2 + 2f'yz + \ldots = 0$, we have $B'C' - F'^2 = a'\Delta'$, $G'H' - A'F' = f'\Delta'$, etc.; then writing, as usual, $A = bc - f^2$, $F = gh - af$, etc., the condition can also be written $\frac{1}{4}(\Theta')^2 = \Delta'\Theta$, the conic which is triangularly inscribed to the other being Σ', or S'.

It is to be noticed that if we take a conic by which S is reciprocated into Σ', a triad of points of S which is self-polar in regard to Σ' becomes a triad of tangents of Σ' which is self-polar in regard to S, so that the condition $\Theta' = 0$ involves two geometrical facts; but a triad of points of S whose joins touch Σ', becomes, when the reciprocating conic is properly chosen, the triad of these three joins. (Cf. Chap. I, Ex. 24.) This is brought out by writing the condition $\Theta'^2 = 4\Delta'\Theta$ symmetrically in regard to the point equation of S and the tangential equation of Σ', as we have done. It may also be remarked, in regard to the two theorems, that the points in which a tangent line of Σ' meets S are conjugate points in regard to the conic whose equation is $2F - S\Theta' = 0$, where $F = 0$ is the conic locus of points from which the tangents to S and Σ' form a harmonic pencil (above, Ex. 17).

To prove that the algebraic conditions above given are necessary and sufficient, is easy, if we remark that they are both expressible by the invariants referred to in Ex. 23; it is necessary then only to verify them for any conveniently chosen triad of points of reference for the coordinates, that is, for any two conveniently chosen forms to which the equations of the conics can be simultaneously reduced.

Thus, for the first condition, suppose that a triad of points, A, B, C, lying on S, is self-polar in regard to Σ'. We can then, referring coordinates to A, B, C, suppose $S = 2fyz + 2gzx + 2hxy$, and $\Sigma' = b'c'u^2 + c'a'v^2 + a'b'w^2$. For these two we evidently have $\Theta' = 0$, which is thus *necessary*. Conversely, suppose $\Theta' = 0$; take the polar, in regard to Σ', of any point, A, of S, and let this meet S in B and C. In regard to A, B, C we can then suppose, as before, that $S = 2fyz + 2gzx + 2hxy$. The tangential equation $\Sigma' = 0$ must be such that the pole of the line BC, in regard to Σ', which in general is the point whose coordinates are A', H', G' (being the point whose equation is $u'(A'u + H'v + G'w) + v'(H'u + \ldots) + w'(G'u + \ldots) = 0$, when (u', v', w') are the coordinates, $(1, 0, 0)$, of the line BC), must, in this case, be the point A, or $(1, 0, 0)$; thus $G' = 0$ and $H' = 0$. The general form of Θ' is however, here, $2fF' + 2gG' + 2hH'$; thus, as we are assuming $\Theta' = 0$, we must have $fF' = 0$. We can only have $f = 0$ if the conic S reduces to the line pair $2x(gz + hy) = 0$, which we exclude. Wherefore $F' = 0$, and Σ' has the form

$$A'u^2 + B'v^2 + C'w^2 = 0.$$

Therefore $\Theta' = 0$ is *sufficient* for the triad A, B, C to be self-polar in regard to Σ'.

In regard to this relation of two conics, we may remark, further, that we have shewn (Chap. I, Ex. 11) that when S is outpolar to Σ', there can be found, on S, sets of four points, A, B, C, D, such that the pair of lines AD, BC are conjugate in regard to Σ', as also are BD, CA and CD, AB. In general, if, from three points A, B, C, in regard to which the equation of Σ' is $A'u^2 + 2F'vw + \ldots = 0$, there be drawn lines AD, BD, CD respectively conjugate to BC, CA, AB, with respect to Σ', they will meet in a point, D, of coordinates x, y, z, given by $xF' = yG' = zH'$; if the points A, B, C be on S, which then has an equation of the form $2fyz + 2gzx + 2hxy = 0$, the condition that this point D should also be on S is $2fF' + 2gG' + 2hH' = 0$, which is exactly $\Theta' = 0$. The equation $\Sigma' = 0$ can be written in the form

$$\frac{1}{F'}(G'H' - A'F')u^2 + \frac{1}{G'}(H'F' - B'G')v^2 + \frac{1}{H'}(F'G' - C'H')w^2$$
$$- F'G'H'\left(\frac{u}{F'} + \frac{v}{G'} + \frac{w}{H'}\right)^2 = 0,$$

and $u = 0$, $v = 0$, $w = 0$, $(F')^{-1}u + (G')^{-1}v + (H')^{-1}w = 0$ are the respective equations of the points A, B, C, D. In general, if (x_1, y_1, z_1), (x_2, y_2, z_2), (x_3, y_3, z_3), (x_4, y_4, z_4) be any four points, and A_1, A_2, A_3, A_4 be any constants, the equation

$$\Sigma A_r(ux_r + vy_r + wz_r)^2 = 0,$$

containing four terms, for $r = 1, 2, 3, 4$, is easily seen to be the tangential equation of a conic in regard to which the line joining any two of the four points is conjugate to the line joining the other two. Dually, sets of four tangents of Σ' exist, whose point pair intersections are conjugate in regard to S. (Cf. p. 218.)

The particular cases of the relation referred to in Ex. 13 of Chap. I are easy to verify. In particular, a single point, (ξ, η, ζ), regarded tangentially as a coincident point pair, of equation $(u\xi + v\eta + w\zeta)^2 = 0$, is inpolar to S, if (ξ, η, ζ) lie thereon; this suggests immediately the form above given for Σ' in terms of the equations of four points of S, if we notice that, when two or more conics, $\Sigma_1' = 0$, $\Sigma_2' = 0$, ... are all inpolar to S, so also is any conic $A_1\Sigma_1' + A_2\Sigma_2' + \ldots = 0$. We easily verify that the equation

$$(u\xi + v\eta + w\zeta)^2 = 0,$$

if $\xi = m_1n_2 - m_2n_1$, $\eta = n_1l_2 - n_2l_1$, $\zeta = l_1m_2 - l_2m_1$, is that formed, by the rule, as the tangential equation of the line pair

$$(l_1x + m_1y + n_1z)(l_2x + m_2y + n_2z) = 0,$$

consisting of any two lines meeting in (ξ, η, ζ).

If $\Sigma_1 = 0$, $\Sigma_2 = 0$, ..., $\Sigma_5 = 0$ be any five conics, a conic $S = 0$ can be found which is outpolar to all of them, this requiring five linear equations for the six coefficients of S. Dually, a conic can be found inpolar to any five given conics. As an application of this, we can prove that, if $S = 0$, $T = 0$, $U = 0$ be the point equations of any three conics, all the conics given by the equation $\theta^2 S + 2\theta T + U = 0$, for varying values of θ, are inpolar to another conic; for the tangential equation corresponding to this is of the fourth degree in θ, and we can find a conic outpolar to the five conics whose tangential equations are those obtained by equating to zero the coefficients of all the powers (θ^r, for $r = 4, 3, 2, 1, 0$) of θ occurring in this tangential equation. This outpolar conic will, we have seen, contain the point of intersection of every line pair contained, for a proper value of θ, in the system $\theta^2 S + 2\theta T + U = 0$. The condition for a conic of this form to be a line pair, being of the third degree in the coefficients, is of the sixth degree in θ. If we assume, once more illogically anticipating the point of view explained in the following Chapter IV, that an equation of the sixth degree has six roots among the system of symbols which we are employing, we can then infer that the six line pairs of the system of conics $\theta^2 S + 2\theta T + U = 0$ intersect in six points lying on a conic.

Coming now to the relation between conics of which one is triangularly circumscribed to the other, suppose that three points, A, B, C, of a conic, S, are such that the joins, BC, CA, AB, touch another conic Σ'. Taking coordinates relatively to these points, we may then suppose that

$$S = 2fyz + 2gzx + 2hxy, \quad \Sigma' = 2F'vw + 2G'wu + 2H'uv.$$

Then, the expression $aA' + 2fF' + ...$ reduces to $2fF' + 2gG' + 2hH'$; and the expression

$$(bc - f^2)(B'C' - F'^2) + 2(gh - af)(G'H' - A'F') + ...$$

reduces to $(fF' + gG' + hH')^2$. The condition $\Theta'^2 = 4\Delta'\Theta$ is thus *necessary*. Conversely, if this condition be assumed, and we draw, from an arbitrary point, A, of S, the two tangents to Σ', meeting S again in B and C, and refer the coordinates to A, B, C, we have $S = 2fyz + 2gzx + 2hxy$, $\Sigma' = A'u^2 + 2F'vw + 2G'wu + 2H'uv$; thus the condition leads to $ghA'F' = 0$. We suppose S not to reduce to two lines, and Σ' not to reduce to two points; wherefore none of g, h, F' is zero, and we infer $A' = 0$, which shews that BC touches Σ'; or the condition is *sufficient*.

We have already seen (Chap. I, Ex. 24) that the triads of points of S whose joins touch Σ' are all self-polar in regard to another conic, Ω. When we take

$$S = 2fyz + 2gzx + 2hxy, \quad \text{and} \quad \Sigma' = 2F'vw + 2G'wu + 2H'uv,$$

and take two points of S of coordinates x_1, y_1, z_1 and x_2, y_2, z_2, the equation of the line joining these is

$$(x_1 x_2)^{-1} f x + (y_1 y_2)^{-1} g y + (z_1 z_2)^{-1} h z = 0,$$

as we verify at once; the condition that this line should touch Σ', which is $f^{-1} F' x_1 x_2 + g^{-1} G' y_1 y_2 + h^{-1} H' z_1 z_2 = 0$, is however also the condition that the points (x_1, y_1, z_1), (x_2, y_2, z_2) should be conjugate in regard to $f^{-1} F' x^2 + g^{-1} G' y^2 + h^{-1} H' z^2 = 0$. It is easily seen that in regard to this conic, S and Σ' are polar reciprocals; this is then the conic Ω. If we take another of the four conics in regard to which S and Σ' are polar reciprocals, say Ω', a triad of points, A, B, C, of S, whose joins touch Σ', reciprocates, in regard to Ω', into a triad of tangents of Σ' whose intersections, say A', B', C', lie on S. The point A' is thus obtained from A by the sequence of two reciprocations, in regard to Ω' and Ω, and, as has been remarked, the line AA' passes through one of the triad of points self-polar in regard to both S and Σ', as also do BB' and CC'. The aggregate of the triads A, B, C may thus be arranged in sets of four, of which any two are in perspective.

If the conics S, Σ' be referred to their common self-polar triad, it is easy to choose a parameter for the points of S, such that the condition, that a chord of S should be a tangent of Σ', takes the form $(\theta^2 \phi^2 + 1) a + \theta \phi b - (\theta^2 + \phi^2) = 0$, and to see (see Ex. 20, above) that the condition that two points, θ, ϕ, so connected should be two of a set of an involution of sets of three points is $a^2 = b + 1$. Introducing this value of b, and adjoining the factor $\theta - \phi$, we obtain, for the cubic equation satisfied by ϕ and the two values, θ_1, θ_2, associated therewith,

$$\sigma^3 - \sigma^2 a \phi (\phi^2 - a) (a \phi^2 - 1)^{-1} - \sigma a + \phi (\phi^2 - a) (a \phi^2 - 1)^{-1} = 0.$$

Thus, in accordance with the theory, the involution of sets of three points is given by $\lambda (a \phi^2 - 1) - \phi (\phi^2 - a) = 0$, for varying values of λ. Four associated sets of this involution are given by λ, $-\lambda$, λ^{-1}, $-\lambda^{-1}$, corresponding to the values ϕ, $-\phi$, ϕ^{-1}, $-\phi^{-1}$, as we easily see. The expression $\Theta'^2 - 4\Delta'\Theta$ can be written as the product of four (irrational) factors; it vanishes when the roots α, β, γ, of the cubic equation $\Delta - \lambda \Theta + \lambda^2 \Theta' - \lambda^3 \Delta' = 0$, are subject to

$$\alpha^{\frac{1}{2}} + \beta^{\frac{1}{2}} + \gamma^{\frac{1}{2}} = 0.$$

Ex. 25. Enumeration of reduced forms for the equations of any two conics.

When two conics intersect in four points which are not distinct, the preceding determination of the equations $a x^2 + b y^2 + c z^2 = 0$, $a' x^2 + b' y^2 + c' z^2 = 0$, fails. When the conics touch in two points, they have an infinite number of common self-polar triads, and their equations can, in an infinite number of ways, be reduced to

$x^2 + y^2 + z^2 = 0$, $x^2 + y^2 + cz^2 = 0$. In this case they can be reciprocated into one another in an infinite number of ways.

Prove that when the conics have one point of contact, their equations can be taken to be $2xy + z^2 = 0$, $2xy + cz^2 + y^2 = 0$. And further that they are, then, polar reciprocals of one another in regard to either of the two conics $2xy \pm c^{\frac{1}{2}}z^2 + \frac{1}{2}y^2 = 0$.

Also, that when three of the points of intersection coincide at one point, their equations can be taken to be

$$2xy + z^2 = 0, \quad 2xy + z^2 + 2yz = 0.$$

And that, then, they are polar reciprocals of one another in regard to $2xy + z^2 + yz - \frac{1}{8}y^2 = 0$.

Finally, when the four intersections coincide, that their equations can be taken to be $2xy + z^2 = 0$, $2xy + z^2 + y^2 = 0$. And that, then, they are polar reciprocals of one another in regard to

$$2xy + z^2 + \frac{1}{2}y^2 - \frac{1}{2}(my + 2z)^2 = 0,$$

for any value of m.

Ex. 26. *A particular involution of sets of three points. Apolar triads.*

It has been seen that an involution of sets of three points on a conic is obtained by the intersections of this conic with a set of conics all passing through a point, O, of the conic and through three other points, A, B, C, not lying on the conic. A special case which is of interest is when A, B, C are a self-polar triad in regard to the given conic. It can then be seen that if the parameters, θ, of two sets of the involution be such that, respectively,

$$a\theta^3 + 3b\theta^2 + 3c\theta + d = 0 \quad \text{and} \quad a'\theta^3 + 3b'\theta^2 + 3c'\theta + d' = 0,$$

then $ad' - 3bc' + 3cb' - da' = 0$. This relation is clearly unaffected if we replace a, b, c, d, respectively, by $a + \lambda a'$, $b + \lambda b'$, $c + \lambda c'$, $d + \lambda d'$, whatever λ may be. It is also unaffected if we use, in place of θ, a parameter, ϕ, given by $(m\theta + n)/(r\theta + s)$; namely, if the cubic equations then become $A\phi^3 + 3B\phi^2 + \ldots = 0$, and $A'\phi^3 + \ldots = 0$, we have $AD' - 3BC' + \ldots = 0$; as may readily be verified. Every two sets of the involution are then apolar triads. Conversely, let P, Q, R and P', Q', R' be any two triads upon a conic; upon the line $Q'R'$ two points, A, C, can be taken which are conjugate to one another in regard to all conics through the four points P', P, Q, R; in particular A, C will then be conjugate in regard to the given conic. If B be the pole of $Q'R'$, in regard to the given conic, and BP' meet this conic again in O, it can be shewn that the six points P, Q, R, A, C, O lie

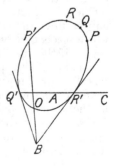

on a conic (above, p. 92). But, when P', Q', R' and P, Q, R are
apolar, this conic also passes through B, as may be proved. The
points A, B, C are a self-polar triad in regard to the given conic;
and there are two conics through these, and through the point O,
whose further intersections with the given conics are, respectively,
the triad P', Q', R' and the triad P, Q, R; namely, the first of these
is the degenerate conic consisting of the lines AC and BO; the
second of these is the conic above remarked. Thus every conic
through A, B, C, O meets the given conic in a triad of points apolar
with P, Q, R, and with P', Q', R'.

If the given conic, referred to Q', R', B, have the equation
$xz - y^2 = 0$, any conic through A, B, C has an equation of the form
$x^2 - \lambda^2 z^2 = 2y\,(fz + hx)$; if this pass through O, or $(m^2, m, 1)$, we
can replace $2f$ by $m^3 - \lambda^2 m^{-1} - 2hm^2$. The three other intersections
of this conic with the given one are then determined by

$$\theta^3 + \theta^2 m + \theta m^2 + \lambda^2 m^{-1} - 2h\theta\,(\theta + m) = 0,$$

which, for every value of h, is apolar with the cubic $\theta^2 + \theta m = 0$,
belonging to Q', R', P'. It is easy to see that A, C are conjugate
in regard to all conics through P', P, Q, R.

If, for $f = ax^3 + 3bx^2y + 3cxy^2 + dy^3$, the Hessian form

$$H = f_{11}f_{22} - f_{12}^2, \text{ where } f_{11} = \partial^2 f/\partial x^2, \text{ etc.,}$$

be denoted by $px^2 + qxy + ry^2$, we at once verify that $ar - bq + cp = 0$
and $br - cq + dp = 0$. This gives the result, previously mentioned
(above, p. 124), that the cubic form $H(Ax + By)$, wherein A, B are
arbitrary, is apolar with f. We have also mentioned (see above,
p. 30) that if the tangents at three points, P, Q, R, of a conic,
meet QR, RP, PQ in D, E, F, the line DEF meets the conic in the
Hessian points, upon the conic, of the points P, Q, R. Wherefore,
if P, Q, R and Q', R' be any points of a conic, the point, P', such
that P', Q', R' are apolar with P, Q, R, may be constructed by
drawing the conic touching QR, RP, PQ, $Q'R'$ and the line DEF;
the tangents to this conic, other than $Q'R'$, drawn from Q' and R',
meet in the point, P', of the given conic, which is required. A
further construction for P' arises from what is said above; we have
only to draw the conic through P, Q, R and B (the pole of $Q'R'$),
which meets $Q'R'$ in points, A, C, conjugate in regard to the given
conic; if this meet the given conic again in O, then P' is on BO.

The line DEF, spoken of above, may, as before (Chap. i, Ex. 9),
be called the Hessian line of P, Q, R. The reader may verify, for
example, that for the three points of a circle, of centre O, consisting
of the absolute points I, J and any point, P, of the circle, the
Hessian line is that through the middle point of O, P, at right
angles to OP. Given three tangents of a conic, we may, dually,
define the Hessian point of these. For example, in the case of a

parabola, of focus S, the Hessian point of the three tangents consisting of SI, SJ and the absolute line IJ, is the point H, such that the directrix is at right angles to SH and passes through the middle point of S, H.

It may also be proved that if two conics, S and S', meet in A, B, C, D, the polar line, in regard to the conic S, of the Hessian point of the tangents of S' drawn at A, B, C, meets S in two points, which, with D, form, on S, a triad apolar with the triad A, B, C. Dually, if the common tangents of two conics, Σ and Σ', be a, b, c, d, and we take the Hessian line of the three points where Σ' touches a, b, c, this Hessian line meets Σ in two points which, taken with the point of contact of d with Σ, form a triad apolar with the points of contact of a, b, c with Σ. (Cf. Chap. i, Ex. 32.)

Ex 27. Three conics related in a particular way.

Suppose that a conic, Σ, touches three lines, BC, CA, AB, respectively at L, M, N; and that H is the Hessian point of the three tangents, being the point common to AL, BM, CN. Further, that

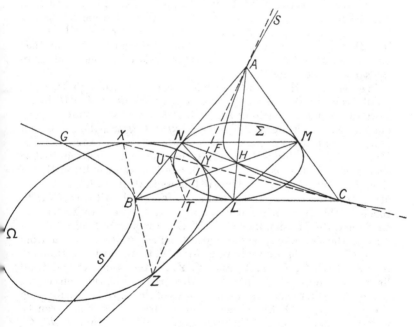

S is any conic, through A, B, C, which passes also through H. Then, for S, the triad L, M, N is self-polar. The tangents of S, at A and C, thus meet on NL, say, in Y, and if these tangents meet BC, BA, respectively, in T and U, the line TU touches Σ. For, taking various conics S, through A, B, C, H, the lines TU envelope a conic, and this is identified with Σ by remarking that particular conics S are the line pairs AB, NC; AL, CB; AC, BH, for which the line TU is, respectively, AB; BC; CA. The line TU is the Hessian line of A, B, C, for the conic S; the relation of the conics S, Σ is thus reciprocal.

The conic S meets MN in two points, F, G, which are harmonic in regard to M and N; thus the point triad A, F, G, of points of S, is self-polar in regard to Σ, and S is outpolar to Σ. If the tangents of S, at A, B, C, meet MN, NL, LM, (in pairs), respectively in X, Y, Z, the line triad YZ, ZX, XY is self-polar for all conics which touch BC, CA, AB and TU, in particular for Σ. Thus S is also inpolar to Σ.

Further, if we consider the locus of the poles, in regard to Σ, of the tangents of S, that is, the polar reciprocal of S in regard to Σ, whose tangents are the polars, in regard to Σ, of the corresponding points of S, we obtain a third conic, say, Ω. As the polar of XZ, in regard to Σ, is the point Y, and the polar of B, in regard to Σ, is NL, with a similar remark for ZY and XY, it appears that Ω touches MN, NL, LM, respectively at X, Y and Z. It will appear that Ω is equally the polar reciprocal of Σ in regard to S; and that any two of the three conics, S, Σ, Ω are related to one another in the same way as were S and Σ.

We see that the condition that a conic, S, through A, B, C, should contain the Hessian point of the lines BC, CA, AB, in regard to a conic Σ touching BC, CA, AB, is that S should be outpolar to Σ. With the symbols, if S be $2fyz + 2gzx + 2hxy = 0$, and Σ be, tangentially, $vw + wu + uv = 0$, which is not a restriction, the Hessian point referred to is $(1, 1, 1)$, and the condition that S should contain this is $f + g + h = 0$, which is the condition that S should be outpolar to Σ. Thus it follows that, if A', B', C' be any other triad of points of the conic S, for which $B'C', C'A', A'B'$ touch Σ, the Hessian point of these lines, in regard to Σ, also lies on S; and the Hessian line of A', B', C', in regard to S, also touches Σ; and the figure may be developed from A', B', C' as it was from A, B, C. In the figure we have considered, there are three triads of points; A, B, C; L, M, N, and X, Y, Z; the points of one triad lie on the joins of the points of another triad, while the joins of the points of the first triad contain the points of the third triad. Keeping the conics, S, Σ, Ω, fixed, such a set of three point triads can be initiated from any one of the nine points; and the triad, A', B', C', of points of S, belonging to such a set, will be apolar with A, B, C. This follows from what has preceded.

The conics S, Σ, being such that S is both outpolar to Σ, and triangularly circumscribed to Σ, are such that the two invariants

Θ, Θ' for these two conics (above described, Ex. 23) are both zero. For we have both $\Theta' = 0$ and $\Theta'^2 = 4\Delta'\Theta$. Referring S and Σ to their common self-polar triad, we find that these, and the conic Ω, may be taken to have equations $x^2 + y^2 + z^2 = 0$, $x^2 + \omega y^2 + \omega^2 z^2 = 0$, $x^2 + \omega^2 y^2 + \omega z^2 = 0$, where ω is such that $(2\omega + 1)^2 + 3 = 0$. Another form to which three such conics can be reduced is $x^2 + 2yz = 0$, $y^2 + 2zx = 0$, $z^2 + 2xy = 0$ (see the concluding problems on Analytical Conics in Wolstenholme's problems).

Another deduction of the three conics may be made by taking three points, A, B, C, of the conic S, then drawing, through the pole, Y, of AC, any line to meet BA, BC, respectively, in N and L, and then describing the unique proper conic, Σ, which touches BA, BC, respectively, at N and L, and also touches AC. If this conic touch AC at M, the conic Ω then arises as touching NL at Y, and touching MN and ML where these are met by the tangent at B of the conic S.

Particular figures in which three such conics arise may be referred to: (1) If O, P, Q, R be the feet of the four normals drawn to a conic, S, from any point, T, and, the point O being kept fixed, the point T be allowed to vary on the normal at O, the triads of lines, QR, RP, PQ, so obtained, are all tangents to a parabola, Σ, and the poles of these lines, in regard to S, describe a rectangular hyperbola, Ω. These three conics are related as above. Their respective equations, that of Σ being tangential, may be taken to be

$$a^{-1}x^2 + b^{-1}y^2 - z^2 = 0, \quad z_0 uv - b^{-1}y_0 uw - a^{-1}x_0 vw = 0,$$

$$xyz_0 + xzy_0 + yzx_0 = 0,$$

where $a^{-1}x_0^2 + b^{-1}y_0^2 - z_0^2 = 0$. (2) If a circle, S, be described to pass through the focus of a parabola, Σ, with its centre on the directrix of the parabola, there is associated with these a rectangular hyperbola, Ω. The equations may be taken to be

$$(x + az)^2 + y^2 - 4a^2 z^2 = 0, \quad y^2 - 4axz = 0, \quad x^2 - y^2 - 3a^2 z^2 - 2axz = 0,$$

the tangential equation of Ω being $-3l^2 + 4m^2 + a^{-2}n^2 + 2a^{-1}nl = 0$. In connexion with this set we may deduce the result that when a parabola touches the lines BC, CA, AB, in such a way that its directrix contains the centre of the circle A, B, C, the nine points circle of the triad which is formed by the intersections of the tangents of the circle drawn at A, B, C, touches this circle at the focus of the parabola.

Ex. 28. *The director circle of a conic touching three lines is cut at right angles by a circle, for which the three lines form a self-polar triad.* Let a, b, c be the lines, Σ the conic touching them, and C the circle. As Σ is inpolar to C, therefore C is outpolar to Σ; thus, by Gaskin's theorem, C cuts the director circle of Σ at right angles.

Ex. 29. *A transformation of conics through four points into conics having double contact.* The equations

$$\xi^{-1}(xy - z^2) = \eta^{-1}y(x - y) = -\zeta^{-1}z(x - y)$$

lead to $\qquad x^{-1}(\xi\eta - \zeta^2) = y^{-1}\eta(\xi - \eta) = -z^{-1}\zeta(\xi - \eta)$

and transform conics, of equation $y^2 - z^2 + \lambda(z^2 - x^2) = 0$, passing through the four points $(1, \pm 1, \pm 1)$, into conics, of equation $(x - y)^2 + \lambda(z^2 - x^2) = 0$, all having double contact on $x - y = 0$.

CHAPTER IV

RESTRICTION OF THE ALGEBRAIC SYMBOLS. THE DISTINCTION OF REAL AND IMAGINARY ELEMENTS

Review of the development of the argument. In the first three chapters of this volume we have followed the plan, adopted in Chapter I of Volume I, of avoiding reference to the relative order of the points of a line, or to characteristics of the algebraic symbols which would follow from such order, save for the occasional anticipation, in some of the latter examples in Chapter III, involved in referring to the roots of an equation. Our discussion of the Real Geometry, in Chapter II of Volume I, led us to regard the figures of such a real geometry, when amplified by means of the postulated elements, as existing among those of the Abstract Geometry. Thus we came to recognise, upon a line of which three points, O, E, U, are given, the existence of an aggregate of points satisfying the conditions for an abstract order (Vol. I, pp. 121, 128); while, however, in the real geometry, two points determined a segment, *two* segments were determined in the abstract geometry by the assignment of two points; and, instead of the possibility of speaking of a point as lying between two others, we were led to speak of the separation of two points of a line by two others (*ibid.* pp. 118, 119). At the same time, the points of a line obtainable, from three given points of the line, by a finite number of repetitions of the process of forming the fourth harmonic point from three given points, were associated, relatively to the three fundamental points, with certain iterative symbols, analogous to the integers of ordinary arithmetic (*ibid.* pp. 86, 138); and it was found, on the basis of the conception of the separation of two points by two others, that the points obtainable by this harmonic process arise in every segment of the line (*ibid.* p. 132). It was suggested (*ibid.* p. 146) that a consequence of the assumption of the existence of an aggregate of points upon the line satisfying the condition (3) for an abstract order (*ibid.* p. 122), would be the introduction of algebraic symbols forming a natural extension of the iterative symbols, as the irrational numbers of arithmetic form an extension of the rational numbers; it would follow from the equable distribution of the points of the harmonic net, that points for which such extended symbols are appropriate would be equably distributed. The deduction of Pappus' theorem (*ibid.* p. 129) was for points satisfying the conditions for an abstract order, it being assumed that the lines involved contained all the points necessary to the satisfaction of these conditions. More fully, the geometrical scheme was built up by supposing certain points to be given, and other points, in infinite number, to be deduced from these by prescribed rules of construction; it was assumed that if the fundamental points were among those for which the conditions for an abstract order, on any lines containing them, were satisfied, the points derived therefrom by the prescribed conditions might equally be regarded as satisfying such conditions. This underlying assumption of the existence of what, relatively to the rules of construction, may be called a *closed* set of points, which is to be made more precise with the use of the algebraic symbols, was avoided, in the Abstract Geometry of Chapter i of Vol. i, by the frank assumption of Pappus' theorem.

We were not content, however, with only points satisfying the conditions for an abstract order. We recognised, also, the existence of points such as would arise if the algebraic symbols appropriate to the geometry were such that the equation $z^2 = c$ could always be satisfied by a symbol, z, of the system employed, whatever symbol of this system c might be. When, in particular, c is -1, this condition requires the existence of points not previously recognised (see Vol. i, p. 162). We shall, as is customary, use i to denote a symbol for which $i^2 = -1$. By the adjunction of this symbol, i, to the symbols previously introduced on the basis of the iterative symbols, it is then possible to obtain a solution of the equation $z^2 = c$ for every symbol c which is itself formed from i and the previous symbols. If we call the symbols consisting of the iterative symbols, and those derived from them as the irrational numbers of arithmetic are derived from the rational numbers, the *real* symbols, all the symbols arising in consequence of the fundamental geometrical constructions, and the possibility of Steiner's problem (Vol. i, p. 155), in virtue of the laws of combination of the symbols (Vol. i, p. 64), will evidently be of the form $x + iy$, where x and y are real symbols.

It is the case now that, if we introduce certain laws of order of succession, or, if the phraseology is allowed, certain conceptions of *magnitude*, the real symbols are, in manipulation, indistinguishable from the real numbers of arithmetic, and the symbols $x + iy$ are indistinguishable from the numbers of ordinary Analysis. To those who have considered the theoretical basis of Analysis, this conclusion will be evident. But this basis is often discussed with the help of conceptions derived from the consideration of the measurement of quantity, and with a metrical Euclidean plane[1]. Some brief further indication of the ideas involved may therefore be allowed.

[1] Cf. Gauss, *Werke*, iii, 1866, p. 79. Ich werde die Beweisführung in einer der Geometrie der Lage entnommenen Einkleidung darstellen, weil jene dadurch die grösste Anschaulichkeit und Einfachheit gewinnt. Im Grunde gehört aber der

To some it will appear that it would have been better, as it would have been simpler, frankly to assume from the first that all points to be considered in the geometry are such as those for which the numbers of Analysis, $x + iy$, are appropriate. To the writer this attitude appears to beg one of the main, and most interesting, questions arising in discussing the foundations of geometry. It is admitted that, by help of the symbols, we can define a closed system of points and figures to which this closed system of symbols is appropriate. But this does not appear to preclude the existence of other points and figures susceptible of geometrical theory. It is true of course that in the present volumes we do not obtain any geometrical results for which corresponding results, identical in statement, do not hold in the realm of points representable by symbols $x + iy$. But this appears to be a consequence of the arbitrary limitation, adopted for practical reasons, involved in the assumption, first of Pappus' theorem, and then of the possibility of Steiner's construction, or, more generally, as in some examples in Chap. III, of the possibility of the solution of problems depending, when expressed algebraically, on equations of higher than the second order. Nor, even, is it the case that the rules for the order of succession of the real symbols, to which we presently proceed, are the only ones we might adopt. For instance, an interesting theory arises by considering a system of symbols representable in the form $a + bx + cx^2 + ...$, wherein a, b, c, ... may be supposed to be real symbols, but x *not* one of these, and agreeing that such a symbol shall be held to follow another symbol of the same form, when, in the difference of these, $a - a' + (b - b')x + (c - c')x^2 + ...$, the first of the coefficients $a - a'$, $b - b'$, $c - c'$, ..., which is not zero, is *positive*.

The distinction of positive and negative for the real symbols. The real symbol, a, of a point, $O + aU$, belonging to the ordered aggregate which we may suppose to exist upon a line, upon which three points of the aggregate, O, U, and E, of symbol $E = O + U$, are given, is said to be *positive* when the point lies in the segment OU which contains E. When the point belongs to the other segment OU, the symbol a is said to be negative. We saw that the points $O + aU$, $O - aU$ are harmonic conjugates in regard to O and U, and, therefore, separated by O and U (Vol. I, pp. 74, 119); thus, when a is a positive symbol, the symbol $-a$ is negative. In particular, as $0 = -0$, it is immaterial whether we regard the symbol of the point O as positive or negative; the point U is exceptional, as will appear, and our definition does not apply to it.

eigentliche Inhalt der ganzen Argumentation einem höhern von räumlichen unabhängigen Gebiete der allgemeinen abstracten Grössenlehre an, dessen Gegenstand die nach der Stetigkeit zusammenhängenden Grössencombinationen sind, einem Gebiete, welches zur Zeit noch wenig angebauet ist, und in welchem man sich auch nicht bewegen kann ohne eine von räumlichen Bildern entlehnte Sprache.

Whether c be a positive or negative real symbol, the symbol c^2 is positive; in virtue of $(-c)(-c) = c^2$ (Vol. i, p. 85), it is sufficient to prove this when c itself is positive. To pass from $P, = O + cU$, to $Q, = O + c^2U$, the construc- tion given (Vol. i, p. 84) was as follows: draw, from O and U, respectively, two arbitrary lines OK, UK, meeting in K; take Y arbitrarily on UK; let EY meet OK in E', and $E'P$ meet KU in X; then, if PY meet OK in P', the point Q is the intersection of $P'X$ with OU. Conversely, if

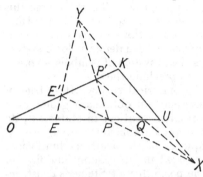

$$Q = O + aU,$$

and a be a positive real symbol, there is one positive real z for which $z^2 = a$, as well as a negative real $-z$. To obtain P from Q is, in fact, the problem considered in Vol. i, p. 157 (iii): we are to draw through the given points E', Q, Y, respectively in the segments $OE'K$, OQU, $UXYK$, three lines, $E'PX$, XQP', $YP'P$, whose in- tersections in pairs, respectively P', P and X, shall lie respectively in the specified segments.

If a, b be both positive real symbols, the symbol $c, = a + b$, is also positive; if a, b be both negative, then c is also negative. In each

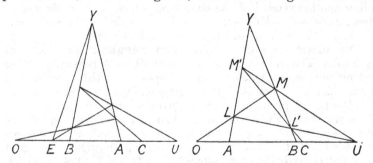

case, the equation $a + b = 0$ requires that both $a = 0$ and $b = 0$. The product ab is, likewise, positive when a, b are both positive, or both negative.

Order of magnitude for the real symbols. The real symbols may then be put in order, which we may call the order of magni- tude, by the rule that $a > b$ when $a - b$ is a positive symbol. If $A = O + aU$, $B = O + bU$, it can be proved that the geometrical condition for this is that the order U, B, A shall be the same as that of O, E, U. In this definition it is understood that neither A nor B is at U. It follows that if x be any real symbol, and $a > b$, then $a + x > b + x$; also if a, b, x, y be positive real symbols, and $a > b$, $x > y$, then $ax > by$. And so on.

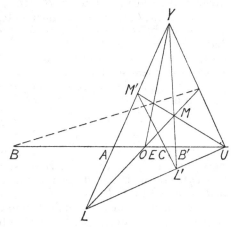

With these definitions it is at once possible to pass to the notion of the upper (or lower) *bound*, of any indefinitely continued sequence of real symbols, when such exists; and to the notion of a continuous function, and the theory of such functions.

The roots of a polynomial. It requires then only an elaboration of these conceptions to shew that a polynomial

$$z^n + (a_1 + ib_1) z^{n-1} + \ldots + a_n + ib_n,$$

wherein $a_1, b_1, \ldots, a_n, b_n$ are real symbols, can be written as the product of n factors of the form $z + p + iq$, where p, q are real symbols.

Writing for z the form $x + iy$, and collecting the real symbols together, the polynomial takes the form $u + iv$, when u and v are polynomials in x and y with only real symbols as coefficients. A proof of the theorem may be constructed by proving that $u^2 + v^2$, which is a real continuous function of x and y, which is necessarily positive, has zero for its least value. In particular, when all of $b_1, b_2, \ldots, b_n$ are zero, if there is a root $x + iy$, with y not zero, there is also a root $x - iy$, so that, when n is odd, there is at least one root, x, which is real.

The exponential and logarithmic functions. We also assume the existence of the so-called exponential function, $\exp(z)$, or e^z, which is unaltered when z is replaced by $z + 2\pi i$, and the functions

$$\cos z, = \tfrac{1}{2} (e^{iz} + e^{-iz}), \quad \sin z, = \frac{1}{2i} (e^{iz} - e^{-iz}),$$

$$\cosh z = \tfrac{1}{2} (e^z + e^{-z}), \quad \sinh z = \tfrac{1}{2} (e^z - e^{-z}).$$

When $e^z = \zeta$, we write $z = \log \zeta$, the function $\log \zeta$ being indeterminate by additive integer multiples of $2\pi i$. The fundamental properties of this function are also assumed.

A simple application to the double points of an involution. If O, U and A, B be two pairs of points, in line, there is a pair of points, P, Q, of the line, which are harmonic both in regard to O, U and in regard to A, B. If, beside O, U, there be given another point, E, of the line, which, in terms of the symbols of O, U, has the symbol $E = O + U$, then the symbols of A, B, and of P, Q, have definite expressions in terms of O and U. It is of importance to know when the expressions of P and Q, in terms of O and U, contain only real symbols.

Let the expressions of A and B be $O + aU$, $O + bU$, respectively; let p, q be two symbols such that $p^2 = a$ and $q^2 = b$; we clearly have

$$q\,(O + p^2 U) + p\,(O + q^2 U) = (p + q)\,(O + pqU),$$
$$- q\,(O + p^2U) + p\,(O + q^2 U) = (p - q)\,(O - pqU);$$

comparing these, remembering that points of symbols $A \pm mB$ are harmonic in regard to A and B, we see that the points $O + pqU$, $O - pqU$ are the points desired, harmonic both in regard to O, U, and in regard to A, B.

Now, when a, b are real, and, therefore, ab is real, the symbol pq, which is such that $(pq)^2 = ab$, is real only when ab is positive, that is, only when a, b are both positive, or both negative. Thus, when O, U, E and A, B are regarded as real, *the points P, Q*, which are harmonic both in regard to O, U and A, B, *are real provided A and B are not separated by O and U, but not otherwise.*

If, however, relatively to O, U, and E, the points A, B are such that $a = x + iy$, and $b = x - iy$, where x, y are real, then ab, which is $x^2 + y^2$, is real and positive; in this case, then, P, Q are real. Two such points, A, B, are commonly said to be *conjugate imaginaries* in regard to O, U and E.

As the results are of frequent application, we may regard the matter from another point of view. Let us refer both O, U and A, B to two other points of the line, say, X and Y, which we suppose to be given, together with another point of the line whose symbol is $X + Y$; suppose then that $O = X + \xi Y$, $U = X + \eta Y$, where ξ, η are the roots of the equation $ax^2 + 2hx + b = 0$, and also that $A = X + \alpha Y$, $B = X + \beta Y$, where α, β are the roots of the equation $a'x^2 + 2h'x + b' = 0$. The general pair of points, of the involution determined by the two pairs O, U and X, Y, is then $X + mY$, $X + nY$, where m, n are the roots of the equation

$$ax^2 + 2hx + b + \lambda\,(a'x^2 + 2h'x + b') = 0,$$

for a proper value of λ. If the symbols ξ, η be either real, or conjugate imaginaries, so that $\xi + \eta$ and $\xi\eta$ are real, we may suppose a, h, b to be real, the case when ξ, η are real being that in which $h^2 - ab$ is positive. Conversely, when a, h, b are real, ξ and η are either real or conjugate imaginaries. Similarly for α and β. We limit ourselves now to the consideration of the case when a, h, b, a', h', b' are all real. The points P, Q, which are harmonic both in regard to

O, U and to A, B, are the double points of the involution determined by these two pairs. These arise, then, for the values of λ which make the last written quadratic a perfect square, namely when $(a + \lambda a') (b + \lambda b') = (h + \lambda h')^2$; the values of λ satisfying this will either be real, or be conjugate imaginaries. If these values be λ_1 and λ_2, the last written quadratic will be the square of

$$(a + \lambda_1 a') x + h + \lambda_1 h', \text{ or of } (a + \lambda_2 a') x + h + \lambda_2 h',$$

and the symbols of P, Q will be

$$(a + \lambda_1 a') X - (h + \lambda_1 h') Y, \quad (a + \lambda_2 a') X - (h + \lambda_2 h') Y;$$

these points P, Q will then be real in regard to X, Y and $X + Y$, if, and only if, λ_1 and λ_2 be real, for which the condition is that

$$(2hh' - ab - a'b)^2 - 4 (h^2 - ab) (h'^2 - a'b')$$

should be positive. If we put

$$M = - a^{-2} (h^2 - ab), \quad M' = - (a')^{-2} (h'^2 - a'b'),$$

this expression is equal to $a^2 a'^2$ multiplied into

$$[(a^{-1}h - (a')^{-1}h')^2 + M - M']^2 + 4M' (a^{-1}h - (a')^{-1}h')^2.$$

We see thus that the expression is positive if either M or M' be positive; that is, if either $h^2 - ab$ or $h'^2 - a'b'$ be negative; that is, if one, or both, of the two pairs O, U and A, B, be conjugate imaginaries in regard to X, Y and $X + Y$. But, when M, M' are both negative, so that ξ, η, α, β are all real, or O, U and A, B are all real in regard to X, Y and $X + Y$, the expression is equal to $a^2 a'^2$ multiplied into $(\xi - \alpha) (\xi - \beta) (\eta - \alpha) (\eta - \beta)$; it is, therefore, only positive when O, U are not separated by A and B.

We may interpret ξ, η and α, β as values of the parameter, θ, on the conic expressed by $x = \theta^2$, $y = \theta$, $z = 1$, speaking of points given by real values of θ as real points. Then the double points of the involution given by the two pairs of points of the conic, O, U and A, B, are the points of contact of the tangents to the conic which can be drawn from the point in which the chords OU and AB intersect. The previous remarks, therefore, give the result that, from any point of a line which does not cut the conic in two real points, there can be drawn two real tangents of the conic. Two real tangents can also be drawn from a point which is such that, from it, two lines can be drawn whose respective pairs of intersections with the conic are both real and do not separate one another, regarded as points of the aggregate of real points of the conic which is given by the real values of θ. But from a point from which two lines can be drawn intersecting the conic each in a pair of real points, those of one pair separating those of the other pair, no real tangents can be drawn to the conic.

We may interpret the result somewhat differently, regarding O, U

as the double points of one involution, and A, B as the double points of another involution; then P, Q are the pair of points common to both involutions. These are then real when at least one of the involutions has for its double points a pair of conjugate imaginary points. Regarding the points, as before, as lying on the conic $xz - y^2 = 0$, the result shews that, if T and T' be two points, not lying on the conic, one of which, at least, is such that no real tangents can be drawn from it to the conic, then the line TT' meets the conic in two real points. For, the involution of which the double points are the points of contact of tangents from T to the conic, consists of pairs of points of the conic lying on lines through T.

Another application is as follows : With reference to two absolute points, I and J, we have defined the axes of a conic, whose centre is Y (the pole of the line IJ), as the lines joining Y to the points, P, Q, of the line IJ, which is harmonic, both in regard to I, J, and in regard to the two points, X and Z, in which the conic meets the line IJ. Then, when, with respect to points regarded as real on the line IJ, the points I and J are conjugate imaginaries, the axes of the conic are real, provided the centre, Y, is real, and the points X, Z, where the conic meets the line IJ, are either real or are conjugate imaginaries.

Remark. In what precedes we have dealt with the reality of P, Q in two ways, first considering A, B relatively to O, U, and afterwards considering both O, U and A, B relatively to two other points X, Y. The distinction is not immaterial; if O, U be imaginary in regard to X and Y, a point which is real in regard to X and Y may be imaginary in regard to O and U. For instance, if $\lambda^2 + \mu^2 = 1$, and $O = X + (p + iq)\,Y$, $U = X + (p - iq)\,Y$, then the point given by $O + (\lambda + i\mu)\,U$ is the same as that given by

$$X + [p + \mu^{-1}(1 - \lambda)\,q]\,Y.$$

General statement of the meaning of real and imaginary points in a plane. As a basis for geometry in a plane, we suppose that four points are given, say A, B, C and D. These are regarded as real. By the intersections of the joining lines of pairs of these points, three other points are obtained. By the intersections of the joining lines of the pairs of all the points now obtained, other points are found, and the method may be indefinitely repeated. The derivation may be arranged as a series of applications of the process of finding the fourth harmonic point of three given points of a line, or of the process of finding the fourth harmonic line of three lines which meet in a point. The points and lines so obtained are unlimited in number. Further, with a proper definition of the meaning of the statement, it appears that the points so derived occur everywhere in the plane. This generalises the theorem for the points of a harmonic net upon a line (Vol. I, p. 131). The symbols of the points A, B, C being fixed relatively to one another by the condition that the symbol of the point D is $A + B + C$, the points so derived are in fact all those expressed by symbols of the form $lA \pm mB \pm nC$, where l, m, n are iterative symbols. This has not been formally proved, but seems sufficiently clear from what was said, Vol. I, p. 86 (see also Moebius, *Ges. Werke*, Leipzig, 1885, Vol. I, p. 247). Thus, as in the case of points of a line, in virtue of the notion of an abstract order (Vol. I, p. 128), these points form a net sufficient to identify a multitude of points of the plane, all expressed by symbols $pA + qB + rC$, where p, q, r are real symbols in the sense which has been explained. It is these points, and no others, which we speak of as the points of the plane which are real with respect to A, B, C and D. All other points of the plane, of symbols $xA + yB + zC$, for which one at least of $x^{-1}y$, $x^{-1}z$, $y^{-1}z$, $y^{-1}x$, $z^{-1}x$, $z^{-1}y$ is a complex symbol $(p + iq)$, the others of these being real, we speak of as imaginary or complex points, relatively to A, B, C and D.

This statement is capable of much more minute analysis. For the purposes we have in view here, it is necessary to trace the connexion between the algebraic symbols employed and the numbers of Analysis; it is to this end that the Moebius net is introduced.

Some applications of the definition. Imaginary points and lines. A line which has two real points, say $pA + qB + rC$ and $lA + mB + nC$, has an infinite number of other points which are real, with symbols $(p + \sigma l)\,A + (q + \sigma m)\,B + (r + \sigma n)\,C$, where σ is any real symbol. Similarly, if two lines which meet in a real point, P, have each a further real point, respectively Q and R, then an infinite number of lines can be drawn through P having each an infinite number of real points; they are those, namely, which contain real points of the line QR. A line which has two real points has an equation of the form $ax + by + cz = 0$, where a, b, c are real symbols. Conversely, a line with such an equation contains an infinite number of real points, and may be called a *real line*. Through the common point of two real lines, which is necessarily real, may be drawn an infinite number of real lines. A real line contains, however, an infinite number of imaginary points; and, through a real point can be drawn an infinite number of imaginary lines.

Conversely, an imaginary line contains one, but only one, real point; and, through an imaginary point there passes one, and only one, real line. For, to an imaginary point,

$$(l + ip)\,A + (m + iq)\,B + (n + ir)\,C,$$

where l, m, n, p, q, r are real but $l^{-1}p$, $m^{-1}q$, $n^{-1}r$ not all equal, there corresponds a *conjugate* imaginary point,

$$(l - ip)\,A + (m - iq)\,B + (n - ir)\,C\,;$$

and, similarly, to an imaginary line there corresponds another, the conjugate imaginary line. The join of an imaginary point to its conjugate is a real line through the first point; the intersection of an imaginary line with its conjugate is a real point of the first line. Conversely, a line given by the equation

$$(l + ip)\,x + (m + iq)\,y + (n + ir)\,z = 0,$$

where l, m, n, p, q, r are all real, but $l^{-1}p$, $m^{-1}q$, $n^{-1}r$ are not all equal (since otherwise this would be a real line), contains only the real point for which both the equations $lx + my + nz = 0$, $px + qy + rz = 0$ are satisfied; through this point the conjugate imaginary line passes. Similarly, for an imaginary point.

The reader will compare this theory with that given in Vol. I (p. 166), of aggregates of real elements having the mutual relations of imaginary elements. Consider any imaginary point. We can, conveniently, suppose its symbol to be $3^{\frac{1}{2}}L + iP$, where

$$3^{\frac{1}{2}}L = lA + mB + nC, \quad P = pA + qB + Cr,$$

its conjugate being $3^{\frac{1}{2}}L - iP$. The values $i/3^{\frac{1}{2}}$ and $-i/3^{\frac{1}{2}}$ are the roots of the equation $3x^2 + 1 = 0$. The quadratic $3x^2 + 1$ is the Hessian form of the cubic $(x - s)\,[(x - s)^2 - (1 + 3sx)^2]$, as is easy to verify, the symbol s being arbitrary. The roots of this cubic are s, $(s - 1)/(1 + 3s)$ and $(s + 1)/(1 - 3s)$; if we use $\vartheta(s)$ to denote $(s - 1)/(1 + 3s)$, and $\vartheta^2(s)$ to denote $\vartheta[\vartheta(s)]$, these roots are s, $\vartheta(s)$, $\vartheta^2(s)$; we easily see that $\vartheta^3(s)$, or $\vartheta[\vartheta^2(s)]$, is equal to s. Thus (see Vol. I, p. 173), the theory referred to would represent the given imaginary point, whose symbol is $3^{\frac{1}{2}}L + iP$, by a set of three points taken, in a definite order, on the real line containing the given point, namely the three points whose symbols are $L + sP$, $L + \vartheta(s)\,P$, $L + \vartheta^2(s)\,P$. The conjugate imaginary point, whose symbol is $3^{\frac{1}{2}}L - iP$, would then be represented by the triad, $L + sP$, $L + \vartheta^2(s)P$, $L + \vartheta(s)\,P$, consisting of the same points in a different order. And, s being arbitrary, the same imaginary point would be represented by an infinite number of equivalent triads.

Real and imaginary conics. More care is necessary in defining a real conic than in defining a real line, since a conic can be expressed by an equation in which all the coefficients are real and still not have any real point. For instance, the conic whose equation, relatively to the given real points of the plane, is $x^2 + y^2 + z^2 = 0$, contains no point whose coordinates are all real. A conic can also have two real points, and no other real points; or, can have four real points, and no other real points. Consider, for example, the conic expressed by the equation

$$z^2 + zx\,(l+i) + zy\,(m+i) - xy\,[h^2 - lm + k + i\,(2h - l - m)] = 0,$$

where l, m, h, k are real symbols, i having the usual significance ($i^2 = -1$). This conic contains the two points $(1, 0, 0)$ and $(0, 1, 0)$, which we regard as real. Any other point of real coordinates lying on this conic must be such as to satisfy both the equations

$$z^2 + lzx + mzy - xy\,(h^2 - lm + k) = 0,\quad zx + zy - xy\,(2h - l - m) = 0\;;$$

these, however, by elimination of z, require the equation

$$[(h - l)\,x - (h - m)\,y]^2 + k\,(x + y)^2 = 0,$$

and this can be satisfied by no real values of x and y if k be positive (other than $x = 0$, $y = 0$, which, by recurring to the original equations, does not lead to a point of the conic). When k is negative, there are two values of $x^{-1}y$ which satisfy this equation, each of which leads to a value of $x^{-1}z$. Thus the conic has two, or four, real points, according as k is positive or negative.

In general, the real points of a conic expressed by an equation $(a + ia')\,x^2 + \ldots = 0$, whose coefficients are not real, are the points common to two conics, each with real coefficients, $ax^2 + \ldots = 0$, $a'x^2 + \ldots = 0$. These common points have coordinates which depend upon the roots of an equation of the fourth degree, wherein the coefficients are real; of such an equation, as we have remarked, the real roots are in number, none, two, or four. Thus, a conic whose coefficients are not real possesses none, or two, or four real points, unless it possesses an infinite number (as when it degenerates into two lines of which one is real and the other is not real).

A conic whose equation has real coefficients may have no real point, as we have seen. If, however, such a conic have one real point, and be not a line pair meeting at this point, it has an infinite number of real points; for any real line drawn through the real point meets the conic in another real point. A conic which is a line pair, consisting of two conjugate imaginary lines, contains one real point, the intersection of these, and no other.

Thus we see that we may conveniently speak of a conic which is not a line pair, as real, regarded as a locus of points, when it is expressed by an equation with real coefficients, and possesses one real point. Such a conic has a real tangent at every real point, and is thus real, also, regarded as an envelope. When we speak of a conic as imaginary, it will be desirable to specify whether its equation has real coefficients or not.

It is obvious that a conic which has five real points is a real conic; it may be defined, in the original way, by two related pencils of lines whose centres are at two of these points. But, a real conic possesses imaginary points, and such points may be among five points which are utilised to define such related pencils of lines. It is unnecessary to enter into an examination of the possibilities.

The interior and exterior points of a real conic. In regard to a real conic, supposed not to be a line pair, we may speak of any real point of the plane which does not lie on the conic, as being either *interior*, or *exterior*, to the conic. For the polar line of a real point, say, *P*, in regard to a conic expressed by an equation with real coefficients, is a real line, and will meet the conic either in two real points, or in two conjugate imaginary points; in the latter case the two tangents that can be drawn to the conic from the point *P* are conjugate imaginary lines, and we speak of the point *P* as *interior* to the conic. It was shewn above (p. 160) that every real line drawn through such an interior point meets the conic in two real points. A point, from which two real tangents can be drawn to the real conic, is spoken of as an *exterior* point; from such a point can be drawn some lines which meet the conic in two conjugate imaginary points, and others which meet it in two real points. It can be proved that, if we take two interior points, and consider the two segments of the line which joins these, which are determined by the two real points in which this line meets the conic, then all the real points of one of these segments are interior points. The real points of a plane are thus divided into two categories by means of the real points of a real conic which is not a line pair.

The behaviour for an imaginary conic which is determined by a real equation is different. The tangents drawn to such a conic from any real point are conjugate imaginary, so that, in one sense, any point might be called an interior point; but every real line meets the conic in two conjugate imaginary points.

The reality of the common points, and of the common self-polar triad, of two conics. We consider two conics which are both expressed by equations with real coefficients; one, or both, of the conics may be real. As has been remarked, the determination of the common points of two such conics may be reduced to the solution of an equation of the fourth degree, with real coefficients. To any root of such an equation which is not real there corresponds another root which is the conjugate imaginary of the former; therefore, it may be shewn, to any one of the four common points of the two conics which is not real, there corresponds another common point, conjugate imaginary to the former; and, the line joining these two common points is real. There is, therefore, always, at least one line pair, of two real lines, passing through the four common points of two such conics. If the four common points of the two conics are all real, and different, there are six lines, each joining two of these points, which are all real; in this case the common self-polar triad, of points and lines, constructed from these joins, consists of three real points and three real lines. If the common points of the two conics consist of the two real, and different, points A, B, with the two conjugate imaginary points C and C', the lines AB, CC' are both real, the lines AC, AC' are conjugate imaginaries, and the lines BC, BC' are conjugate imaginaries. The point, P, of intersection of AB and CC' is, therefore, real; the point, Q, of intersection of AC and BC' is conjugate imaginary to the intersection, R, of the lines AC', BC, which are, respectively, conjugate imaginary to AC and BC'; thus the line QR is real, and the common self-polar triad, P, Q, R, consists of one real point and one real line, together with two conjugate imaginary points, and two conjugate imaginary lines. Lastly, if the intersections of the two conics consist of two conjugate imaginary points A, A' and two conjugate imaginary points C, C', the lines AA', CC' are real, and meet in a real point, P; the lines AC, $A'C'$ are conjugate imaginary and meet in a real point, Q, and the lines AC', $A'C$ are conjugate imaginary and meet in a real point, R. The common self-polar triad is thus wholly real. This last case arises, for example, when one of the conics is real, and the other, though having a real equation, is imaginary. Incidentally, we thus give a proof that the cubic equation expressed by

$$\begin{vmatrix} a-\lambda a', & h-\lambda h', & g-\lambda g' \\ h-\lambda h', & b-\lambda b', & f-\lambda f'' \\ g-\lambda g', & f-\lambda f'', & c-\lambda c' \end{vmatrix} = 0,$$

upon which, as we have seen (Chap. III, Ex. 22), the determination of the common self-polar triad of the two conics $ax^2 + \ldots = 0$, $a'x^2 + \ldots = 0$ depends, has real roots when one at least of the two quadric forms occurring on the left in the equations of the conics, is such that it preserves a constant sign for all real values of x, y, z— it being understood that all the coefficients $a, \ldots, a', \ldots$, are real.

The tangential equation of a conic, whose point equation is real, has also real coefficients. We can therefore state the preceding results also in terms of the reality of the common tangents. But the dependence of the reality of the common tangents upon the reality of the common points is a matter also for enquiry.

We have disregarded the possibilities of coincidence of the four common points of the two conics in what has been said; the omission may easily be supplied.

Ex. A conic passes through three points, A, B, C, and has the points D, D' as conjugate points, and the points E, E' as conjugate points. Find two other real points of the conic. In particular, when E, E' are on the line D, D', and separate these.

CHAPTER V

PROPERTIES RELATIVE TO AN ABSOLUTE CONIC. THE NOTION OF DISTANCE. NON-EUCLIDEAN GEOMETRY

The interval of two points of a line relatively to two given points of the line. If M, N, P, Q, be four points of a line, and the symbols of these points be such that $aP = M + xN$, $bQ = M + yN$, then the symbol yx^{-1} is such that: (1), it is unaltered by replacing any one or more of the symbols P, Q, M, N, respectively, by pP, qQ, mM, nN, where p, q, m, n are any symbols of the system employed, of which the commutativity in multiplication is, of course, assumed; (2), it has the same value as for any other four points, M', N', P', Q', which form a range related to M, N, P, Q; with a slight change of notation this is the same statement as that a range of points, $M, N, M + N, M + \lambda N$, is related to a range of points represented by $M', N', M' + N', M' + \lambda N'$. A proof has been given, Vol. I, p. 154; (3), the symbol yx^{-1} is equal to the symbol formed by the analogous rule from the equations which express M, N in terms of P and Q, that is, the equations

$$- a^{-1} (x - y) M = yP - xba^{-1}Q, \quad a^{-1}y (x - y) N = yP - yba^{-1}Q.$$

The symbol can then be represented by $(P, Q; M, N)$, or by $(M, N; P, Q)$. What is of the greatest importance to us here is the fact that, if P, Q, R be any three points of the line MN, we have

$$(P, Q; M, N) . (Q, R; M, N) = (P, R; M, N).$$

Now, and in all that follows, we suppose that the algebraic symbols employed are those arrived at in the preceding chapter, which have the same properties in computation as the complex numbers of ordinary Analysis, with which they may thus be identified. If then $\phi(\xi)$ be any function for which $\phi(\xi\eta) = \phi(\xi) + \phi(\eta)$, and ξ denote $(P, Q; M, N)$, η denote $(Q, R; M, N)$, and ζ denote $(P, R; M, N)$, we shall have $\phi(\zeta) = \phi(\xi) + \phi(\eta)$. We apply this to the case when $\phi(\xi) = (2i)^{-1} \log(\xi)$.

Let the expression

$$\frac{1}{2i} \log [(P, Q; M, N)]$$

be called the *interval* from P to Q with respect to M and N. The interval from P to R is thus the sum of the interval from P to Q and that from Q to R, subject however to the ambiguity inherent in the definition. As $(Q, P; M, N)$ is the inverse of $(P, Q; M, N)$, the interval from Q to P is the negative of the interval from P to Q. But further, in virtue of the property of the logarithm, the interval, in default of further specification, is subject to the addition of an arbitrary integral multiple of π. We may agree that when P and Q coincide, in which case $(P, Q; M, N) = 1$, the interval vanishes, and, then, that when the interval is π, the two points again coincide. We may, however, apply the definition to cases in which the interval is not real, as will be seen. If the interval be denoted by θ, we have $(P, Q; M, N) = e^{2i\theta}$, and, when P, Q, M, N are all real points, it is $i\theta$ which is real. When $\theta = \frac{1}{2}\pi$, we have $(P, Q; M, N) = -1$; then P and Q are harmonic in regard to M and N.

Angular interval in regard to two absolute points. We have considered, in Chapter III, some of the consequences of supposing two points, I, J, of the plane, to be given, and taken as absolute points. With two such points, we may employ the considerations just given, to define the interval from one line, OP, to another line, OQ; namely, if these lines, respectively, meet the line IJ in P' and Q', we may take for this interval the value $(2i)^{-1} \log [(P', Q'; I, J)]$. This depends only on the points, P' and Q', where the lines meet the line IJ, and not on the position of the intersection, O, of the lines. Suppose in particular, when the points of the plane are referred to three points, C, A, B, which we regard as real, that $I = A + iB$, $J = A - iB$, while P', Q' are given by

$$P' = \cos \alpha . A + \sin \alpha . B \quad \text{and} \quad Q' = \cos (\alpha + \theta) . A + \sin (\alpha + \theta) . B;$$

the lines CI, CJ, CP', CQ' can then be expressed by the respective equations $y = ix$, $y = -ix$, $y = \tan \alpha . x$, $y = \tan (\alpha + \theta) . x$. These relations give

$$2e^{i\alpha} P' = I + e^{2i\alpha} J, \quad 2e^{i(\alpha + \theta)} Q' = I + e^{2i(\alpha + \theta)} J$$

and hence $\qquad (P', Q'; I, J) = e^{2i\theta}.$

This equation, approached from a metrical point of view, was, it seems, first stated by Laguerre (*Nouv. Annal.* xii, 1853; *Oeuvres* (1905) II, p. 12). As will be seen, it may be regarded as the foundation of all that follows in this Chapter. (Cp. p. 195.)

It is clear that, if O be any point of a conic containing I and J, of which P and Q are two fixed points, the interval from OP to OQ, in regard to I and J, is the same for all positions of O on the conic. In particular, when PQ passes through the pole of IJ, in regard to his conic, this interval is $\frac{1}{2}\pi$.

Interval in regard to an absolute conic. The postulation of two absolute points thus furnishes a ready means of defining the interval from one line to another. It does not however, except by a limiting process, furnish a means of defining the interval between two points, unless they lie on the line determined by the absolute points. The two absolute points may, however, be thought of as a degenerate conic envelope. This may suggest that we take a fixed undegenerate conic, given in the plane, as a basis for reference. If we do this, two points, say *M* and *N*, are determined on any line, *PQ*, which does not touch the conic, namely the points of inter-section of *PQ* with this absolute conic; and then the interval from *P* to *Q* may be defined, as above, in respect to *M* and *N*. But this conic also determines two lines through an arbitrary point, *O*, of the plane, not lying on the conic, namely the tangents from *O* to this absolute conic; and then the angular interval between any two lines which intersect in *O* may be defined, as above, by means of these tangents. We have thus a method of defining both the interval from one point to another, and the interval from one line to another, in harmony with the duality otherwise existing in the geometry of the plane. When the absolute conic, regarded tan-gentially, becomes a point pair, we thence obtain what has, on other grounds, been chosen as the measure of an angle; we could, similarly, obtain a measure of the interval between two points, if we supposed the existence of two absolute lines in the plane, as a degenerate form of the absolute conic. From this general point of view, the ordinary metric relations, usually employed at least since the time of the Greeks, appear as an unsymmetric and incomplete scheme. Moreover, *distance*, as usually understood, which we shall shew to be deducible from what has been said by a limiting process, so far from being an obvious notion, characterising the space of experi-ence, appears as an artificial, if useful, application of the algebraic symbols.

The idea of length is, presumably, suggested by our experience of what are called rigid bodies. We may refer, at once, here, to the notion which replaces this in the present point of view. Let *P, Q, M, N* be four points in line, of respective coordinates (x, y, z), (x', y', z'), (ξ, η, ζ), (ξ', η', ζ'). Let P_1, Q_1, M_1, N_1 be other four points, the coordinates of P_1 being definite linear functions of the coordinates of *P*, say

$$x_1 = ax + by + cz, \quad y_1 = lx + my + nz, \quad z_1 = px + qy + rz,$$

the coordinates of Q_1 being the *same* linear functions of the coordi-nates of *Q*, those of M_1 the *same* linear functions of the coordinates of *M*, and those of N_1 the same linear functions of the coordinates of *N*. It may then be shewn without difficulty that P_1, Q_1, M_1, N_1 are in line, and are a range which is related to the range *P, Q, M, N*. The linear transformation employed is, in fact, the representation of the correspondence between two related planes (here coinciding) which was considered in Vol. I, p. 149.

Of such linear transformations, depending on eight parameters (the ratios of the coefficients $a, b, c, l, m, \ldots$), there is a set, wherein the eight parameters are such functions of three parameters that any point of an arbitrarily chosen conic gives rise to another point of the same conic; as we shall see in more detail later. Any two transformations of this set, carried out in succession, lead from a point of this conic to another point of this conic, and so, in combination, constitute a transformation of the same set.

If the conic be the absolute conic by which we measure the interval between two points P and Q, as has been explained, M and N being points of the conic, the interval between the transformed points, P_1 and Q_1, has, clearly, from what has been said, the same measure as the interval PQ.

The change, by a linear transformation leaving the absolute conic unaltered, from the interval PQ to the interval P_1Q_1, is the operation which, from our present point of view, replaces what, in the metrical geometry, is described as the movement of PQ into the position P_1Q_1.

The reader may shew that the general linear transformation changing any point of the conic $x^2 + y^2 + z^2 = 0$ into another point of this conic, is given by

$$x_1 = \tfrac{1}{2}(B^2 + C^2 - A^2 - D^2)x + \quad (AB - CD)y + \quad (CA + BD)z,$$
$$y_1 = \quad (AB + CD)y + \tfrac{1}{2}(C^2 + A^2 - B^2 - D^2)y + \quad (BC - AD)z,$$
$$z_1 = \quad (CA - BD)z + \quad (BC + AD)y + \tfrac{1}{2}(A^2 + B^2 - C^2 - D^2)z,$$

where A, B, C, D are arbitrary.

We now consider the matter more in detail: Let the absolute conic be given, with respect to real points of reference for the coordinates, by an equation with real coefficients; it may, thus, be either a real conic or an imaginary conic. We can suppose the points of reference, still being real, to be so chosen that this equation is of the form $z^2 + \kappa(x^2 + y^2) = 0$, it being supposed that this conic is not degenerate. The coefficient, κ, is then real; it is negative for a real conic, but positive for an imaginary conic.

The proof of this form may be given, as follows; First, a homogeneous quadratic form in *two* variables, x, y, with real coefficients, can be reduced, by a real linear change of coordinates, to a sum of squares, with real coefficients. For, let the form be $ax^2 + 2hxy + by^2$. If both a and b be zero, this is $\tfrac{1}{2}h[(x+y)^2 - (x-y)^2]$, which consists of two squares. If not, suppose a not to be zero; then the form is $a(x + a^{-1}hy)^2 + (b - a^{-1}h^2)y^2$, which again consists of two squares. Second, consider the form in *three* variables, consisting of three pairs of terms such as $ax^2 + 2fyz$. If one of the three coefficients a, b, c, say the coefficient a, be not zero, the form is

$$a(x + a^{-1}hy + a^{-1}gz)^2 + (b - a^{-1}h^2)y^2$$
$$+ (c - a^{-1}g^2)z^2 + 2(f - a^{-1}gh)yz;$$

here the three terms after the first are a quadratic form in y and z only, with real coefficients, and can be written as a sum of two (or fewer) squares, with real coefficients. If, however, all of a, b, c be zero, but none of f, g, h be zero, the original form in x, y, z is

$$2fgh(f^{-1}x + \tfrac{1}{2}g^{-1}y + \tfrac{1}{2}h^{-1}z)^2 - 2f^{-1}ghx^2 - \tfrac{1}{2}fgh(g^{-1}y - h^{-1}z)^2,$$

which consists of three squares. If all of a, b, c be zero, and one, or two, of f, g, h be zero, the form reduces to a product of two real linear forms, PQ, which is $\tfrac{1}{4}(P+Q)^2 - \tfrac{1}{4}(P-Q)^2$. In any case, therefore, the form is reducible, in many ways, to $pX^2 + qY^2 + rZ^2$, where p, q, r are real, and X, Y, Z are real linear functions of x, y, z. Assuming that the form, when equated to zero, does not represent a line pair, no one of p, q, r is zero. And, in the equation so obtained, we may suppose that all of p, q, r are positive, or only one is positive. When r is positive, and p, q are also positive, by taking $X_1 = X(p/\kappa)^{\frac{1}{2}}$, $Y_1 = Y(q/\kappa)^{\frac{1}{2}}$, $Z_1 = Zr^{\frac{1}{2}}$, where κ is an arbitrary real positive number, we obtain the equation $Z_1^2 + \kappa(X_1^2 + Y_1^2) = 0$. When r is positive, and p, q are both negative, we can put

$$X_1 = X(-p)^{\frac{1}{2}}/(-\kappa)^{\frac{1}{2}}, \quad Y_1 = Y(-q)^{\frac{1}{2}}/(-\kappa)^{\frac{1}{2}}, \quad Z_1 = Zr^{\frac{1}{2}},$$

where κ is an arbitrary real negative number, and so obtain the equation $Z_1^2 + \kappa(X_1^2 + Y_1^2) = 0$.

Case of an imaginary conic as Absolute. Let (x_0, y_0, z_0) and (x, y, z) be any two points not lying on the absolute conic. Let f denote the expression $z^2 + \kappa(x^2 + y^2)$, and f_0 what this becomes when x_0, y_0, z_0 are put for x, y, z, respectively; also let ψ denote the expression $zz_0 + \kappa(xx_0 + yy_0)$. For the case of an imaginary conic, κ is positive, and f, f_0 are positive; and $ff_0 - \psi^2$, which is equal to

$$\kappa[\kappa(xy_0 - x_0y)^2 + (zx_0 - z_0x)^2 + (zy_0 - z_0y)^2],$$

is also positive. Let M, N denote the points in which the line joining the point (x_0, y_0, z_0) to the point (x, y, z) meets the conic. We can suppose that, for a positive real value of θ, one of these points has the coordinates

$$x_0f_0^{-\frac{1}{2}} - e^{i\theta}xf^{-\frac{1}{2}}, \quad y_0f_0^{-\frac{1}{2}} - e^{i\theta}yf^{-\frac{1}{2}}, \quad z_0f_0^{-\frac{1}{2}} - e^{i\theta}zf^{-\frac{1}{2}},$$

where the square roots, $f_0^{-\frac{1}{2}}, f^{-\frac{1}{2}}$, have their real positive values; for the substitution of these coordinates in the equation of the conic gives the condition

$$1 - 2e^{i\theta}\psi f_0^{-\frac{1}{2}}f^{-\frac{1}{2}} + e^{2i\theta} = 0,$$

namely $$\cos\theta = \psi f_0^{-\frac{1}{2}} f^{-\frac{1}{2}},$$

of which the value is less than unity. If then, relatively to the points of reference for the coordinates, A, B, C, we define the symbols, O and P, of the points (x_0, y_0, z_0) and (x, y, z), respectively, by means of

$$O = f_0^{-\frac{1}{2}} (x_0 A + y_0 B + z_0 C), \quad P = f^{-\frac{1}{2}} (xA + yB + zC),$$

we may write, for the symbols of the points M and N.

$$M = O - e^{-i\theta} P, \quad N = O - e^{i\theta} P,$$

where $\quad \cos\theta = \psi f_0^{-\frac{1}{2}} f^{-\frac{1}{2}}, \quad \sin\theta = (ff_0 - \psi^2)^{\frac{1}{2}} f_0^{-\frac{1}{2}} f^{-\frac{1}{2}},$

and the symbol, $(O, P; M, N)$, defined above (p. 166), is, then, given by

$$(O, P; M, N) = e^{2i\theta}.$$

Now let the point where the line OP meets the polar line of O in regard to the absolute conic be denoted by U. The coordinates (ξ, η, ζ) of this point are given by $\xi = xf_0 - x_0\psi$, $\eta = yf_0 - y_0\psi$, $\zeta = zf_0 - z_0\psi$; for these belong to a point on the line joining (x_0, y_0, z_0) to (x, y, z), and are such that $\zeta z_0 + \kappa(\xi x_0 + \eta y_0) = 0$. With these values we have $\zeta^2 + \kappa(\xi^2 + \eta^2)$ equal to $f_0 (ff_0 - \psi^2)$, which is positive. If we denote the positive square root of this by μ, we have three equations, for x, y, z, of the following form

$$xf^{-\frac{1}{2}} = \psi f_0^{-\frac{1}{2}} f^{-\frac{1}{2}} \cdot x_0 f_0^{-\frac{1}{2}} + (ff_0 - \psi^2)^{\frac{1}{2}} f_0^{-\frac{1}{2}} f^{-\frac{1}{2}} \cdot \xi \mu^{-1}.$$

We may therefore write, using for the point U the symbol $U = \mu^{-1}(\xi A + \eta B + \zeta C)$,

$$P = \cos\theta \cdot O + \sin\theta \cdot U.$$

Conversely, this equation may, if preferred, be used as giving the definition of θ. If we regard O and U as fixed, it gives a real position of P upon the line OU for every real value of θ, all such positions being obtained by restricting θ to the range from 0 to π. In particular, when $\theta = 0$, the point P is at O; and when $\theta = \frac{1}{2}\pi$, the point P is at U, the conjugate point to O in regard to the absolute conic.

By a precisely similar argument any other point Q, of the line OU, can be expressed, in terms of P and the point, P', conjugate to P upon this line, by the equation

$$Q = \cos\phi \cdot P + \sin\phi \cdot P'.$$

As the original formula was equivalent to $(O, P; M, N) = e^{2i\theta}$, so this formula is equivalent to $(P, Q; M, N) = e^{2i\phi}$. Thus we have $(O, Q; M, N) = e^{2i(\theta+\phi)}$, so that, also

$$Q = \cos(\theta + \phi) \cdot O + \sin(\theta + \phi) \cdot U.$$

This, however, is

$$\cos \phi \, (\cos \theta \, . \, O + \sin \theta \, . \, U) + \sin \phi \, (- \sin \theta \, . \, O + \cos \theta \, . \, U) \, ;$$

thus we see that

$$P' = \cos \, (\theta + \tfrac{1}{2}\pi) \, . \, O + \sin \, (\theta + \tfrac{1}{2}\pi) \, . \, U,$$

and, when P is given, in terms of O and U, by θ, then P', the conjugate of P, in regard to the conic, upon the line OU, is given by
$\theta + \tfrac{1}{2}\pi$. This fact can, of course, be directly verified, the coordinates
of P' being, as we easily see,

$$- x_0 \, (ff_0 - \psi^2) + \xi\psi, \quad - y_0 \, (ff_0 - \psi^2) + \eta\psi, \quad - z_0 \, (ff_0 - \psi^2) + \zeta\psi.$$

In words, the interval, in regard to the conic, from a point P to
any point on its polar line, is $\tfrac{1}{2}\pi$.

In a precisely similar way, the interval from a line OA to a line
OB may be defined with reference to the tangents, OM', ON', drawn
from O to the absolute conic. If the points A, B be on the polar of
O, the interval is, by definition, equal to the interval from the point
A to the point B. In particular when OB is conjugate to OA, in
regard to the absolute conic, the interval between these lines is $\tfrac{1}{2}\pi$.

Application to a triangle. We may illustrate these definitions
by applying them to the intervals between any three given points,
not lying in line, and the intervals between the lines which join
these points. But some preliminary remarks will be useful, in view
of the ambiguity we have noticed in the definition of the interval
between two given elements.

Any two points determine *two* segments, in one or other of which
any other real point, of the line determined by the two given points,
must lie. If then we have three points, A, B, C, which are not in
line, and speak of the segments BC, CA, AB, there are eight ways
in which this may be understood. We desire, however, to associate
with each of the segments, say, with BC, which may be regarded as
an aggregate of points, a certain aggregate of lines, all passing
through the point, A, not on the line BC, so that every one of the
lines shall contain one of the points of the segment. Limiting ourselves to *real* points and lines, relatively to A, B, C, if we speak of
this aggregate of lines as an *angle*, we may say that there are, for
each of the points such as A, two angles. We have then in all eight
possibilities, of three pairs, each pair being constituted by a segment
and its corresponding angle. These eight possibilities fall, however,
into two sets, each of four possibilities. In any one of the first set
of four possibilities, the three segments, say, BC, CA, AB, and the
associated angles, say, A, B, C, have, in regard to real points, the
two following properties : (1) Any line of the plane either contains
no point of any of the three segments, or contains a point of each
of two of the segments ; (2) If any point P of the plane be joined

to A, B and C, and we consider the intersections of the joining lines
PA, PB, PC, respectively, with the lines BC, CA, AB, either one,
of these intersections, or all, lie in the segments of these lines which

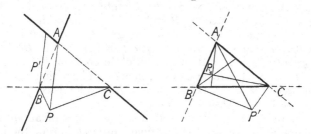

define the possibility considered. If the two segments of the line
BC be denoted by a and a', and the two respectively associated
angles at A by A and A', with a similar notation for B and C, and
one of the four possibilities spoken of consist of A, a; B, b; C, c,
then another of these consists of A, a; B', b'; C', c', and the other
two are respectively A', a'; B, b; C', c' and A', a'; B', b'; C, c.

The other four possibilities consist of the associations, A', a';
B, b; C, c and A', a'; B', b'; C', c', together with A, a; B', b'; C, c

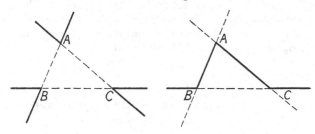

and A, a; B, b; C', c'. For these also a statement can be made as
to the intersections of an arbitrary line with the segments considered, and the lines joining an arbitrary point, to A, B, C, which
belong to the angles associated therewith. But this statement will
be different from that which holds for the first four possibilities.
It appears unnecessary to enter into a detailed analysis of the statement made (cf. Vol. i, p. 100).

In what follows we confine ourselves to one of the first four
possibilities. It is immaterial which one; but we could, if desired,
describe that one by supposing an additional line given in the plane,
whose intersections with the lines BC, CA, AB would enable us to
distinguish between the two segments of the line. If we denote the
intervals BC, CA, AB chosen, measured relatively to the absolute
conic, respectively by a, b, c, the other supplementary segments

BC, *CA*, *AB* may be taken to be, respectively $\pi - a$, $\pi - b$, $\pi - c$, as we have seen. Similarly the interval between *AB* and *AC*, associated with *A*, being denoted by *A*, the supplementary interval between these lines will be $\pi - A$; the intervals adopted at *B* and *C* will likewise be denoted by *B* and *C*. So far, as we have seen, this statement is ambiguous. But if we agree that *a* shall be positive and between 0 and $\frac{1}{2}\pi$, when *the point, B′, of the line BC, which is conjugate to B*, in regard to the absolute conic, *does not lie in the segment BC* which is denoted by *a*, then *a* will be without ambiguity. In this case, the point, *C′*, of the line *BC*, which is conjugate to *C*, in regard to the absolute conic, will also not lie in the chosen segment *BC*; for, as *B, B′* and *C, C′* are pairs of an involution whose double points are the imaginary intersections of the line *BC* with the imaginary absolute conic, the points *B, B′* are separated by the pair *C, C′*. (Above, p. 158.) Similarly we agree that *A* shall be positive and between 0 and $\frac{1}{2}\pi$, when the line through *A* which is conjugate to *AB* does not lie in the angle called *A*, in which case, for a similar reason, the line through *A* conjugate to *AC* will not lie in the chosen angle *A*. And we make the like agreement for *b, c, B* and *C*.

We can, for the sake of algebraic symmetry, suppose, for the present, without loss of generality, that the equation of the absolute conic is $x^2 + y^2 + z^2 = 0$. If then, for example, (x, y, z) and (x', y', z') be the coordinates of the points *A* and *B*, respectively, the interval *c*, of *A, B*, will be such (p. 171, above) that

$$\cos c = \pm \frac{xx' + yy' + zz'}{(fx)^{\frac{1}{2}} (fx')^{\frac{1}{2}}},$$

$$\sin c = \frac{[(yz' - y'z)^2 + (zx' - z'x)^2 + (xy' - x'y)^2]^{\frac{1}{2}}}{(fx)^{\frac{1}{2}} (fx')^{\frac{1}{2}}},$$

where fx denotes $x^2 + y^2 + z^2$, and, for each of the square roots, the positive real value is to be taken. In $\cos c$, the numerator vanishes and changes sign when (x', y', z') passes over the point, of the line joining this to (x, y, z), which is conjugate to (x, y, z). We can, therefore, agree to attach to the coordinates of (x', y', z') such a sign that the numerator in $\cos c$ is positive when *c*, as defined above, is less than $\frac{1}{2}\pi$, and then omit the ambiguous sign $\pm$ in the above formula for $\cos c$.

We now take, first, the particular case in which the lines *AB*, *AC* are conjugate to one another in regard to the absolute conic. We denote the coordinates of the points *A, B, C*, respectively, by (x, y, z), (x', y', z'),

(x'', y'', z''), and denote by Δ the absolute, positive, value of the determinant

$$\begin{vmatrix} x, & y, & z \\ x', & y', & z' \\ x'', & y'', & z'' \end{vmatrix}$$

the minors in this determinant, each with its proper sign, being denoted, respectively, by

$$\begin{matrix} u, & v, & w, \\ u', & v', & w', \\ u'', & v'', & w'' \ ; \end{matrix}$$

any minor of the determinant formed by these last elements is equal to the complementary minor of the original determinant, multiplied by the value of this determinant itself, as is well known. We have the identity

$$(yz' - y'z)^2 + (zx' - z'x)^2 + (xy' - x'y)^2$$
$$= (x^2 + y^2 + z^2)(x'^2 + y'^2 + z'^2) - (xx' + yy' + zz')^2 ;$$

from this, or independently, we have

$$\Sigma (yz' - y'z)(yz'' - y''z) = \Sigma x^2 . \Sigma x'x'' - \Sigma xx' . \Sigma xx'',$$

which we denote by

$$-f(u', u'') = fx . f(x', x'') - f(x, x')f(x, x'') \ \ldots\ldots(i).$$

We similarly have

$$-\Delta^2 f(x', x'') = fu . f(u', u'') - f(u, u')f(u, u'') \ \ldots(ii).$$

Also, as $ux + vy + wz = 0$ is the equation of the line BC, and so on, we have, as equivalent to the condition that AB and AC are conjugate, the equation

$$f(u', u'') = 0 \ \ldots\ldots\ldots\ldots\ldots\ldots\ldots\ldots(iii).$$

We shall write P, P', P'', U, U', U'', respectively, for the real positive numbers $(fx)^{\frac{1}{2}}, (fx')^{\frac{1}{2}}, (fx'')^{\frac{1}{2}}, (fu)^{\frac{1}{2}}, (fu')^{\frac{1}{2}}, (fu'')^{\frac{1}{2}}$.

For the interval c we have, then,

$$\cos c = f(x, x')/PP', \quad \sin c = U''/PP',$$

the $f(x, x')$ being taken positive, or negative, according as $c < \frac{1}{2}\pi$, or $c > \frac{1}{2}\pi$. For the angle C we have, similarly,

$$\cos C = \pm f(u, u')/UU', \quad \sin C = \Delta P''/UU';$$

but, the pole of the line CA is on the line AB, and is the point thereon which is conjugate to A; the line joining C to this point is the line through C which is conjugate to CA. Thus C is greater or less than $\frac{1}{2}\pi$, according as c is greater or less than $\frac{1}{2}\pi$. Also, from equations (ii), (iii), we have $-\Delta^2 f(x, x') = f(u'')f(u, u')$, which

shews that $f(u, u')$ and $f(x, x')$ are of opposite sign. Wherefore we have
$$\cos C = -f(u, u')/UU'.$$

We can now shew that, of the three intervals a, b, c, either all are less than $\frac{1}{2}\pi$, or, just two of them are less than $\frac{1}{2}\pi$. For, let the polar line of the point B meet the line CA in H, so that H will be the pole of the line AB, and will be the point of the line CA which is conjugate to A. By our definition of the triangle considered, if this line, the polar line of B, do not contain a point of either of the segments BA, BC, then H will not be in the segment AC; so that, if each of the intervals a, c is less than $\frac{1}{2}\pi$, then b is, also, less than $\frac{1}{2}\pi$. But, if this line contain a point of each of the segments BA, BC, then, also, H is not in AC; so that, if each of a, c is greater than $\frac{1}{2}\pi$, then, also, b is less than $\frac{1}{2}\pi$. When the polar line of B contains a point of only one of the segments BA, BC, we reach the same conclusion.

But equations (i), (iii) give $f(x', x'')f(x) = f(x, x')f(x, x'')$, which is the same as
$$\frac{f(x', x'')}{P'P''} = \frac{f(x, x')}{PP'} \cdot \frac{f(x, x'')}{PP''};$$

if then we suppose, when x, y, z have been chosen, that the signs of x', y', z' and of x'', y'', z'' are so taken that, not only
$$\cos c = f(x, x')/PP',$$
as has already been agreed, but also $\cos b = f(x, x'')/PP''$, we can infer, remembering that, as has just been seen, one or all of a, b, c are less than $\frac{1}{2}\pi$, that
$$\cos a = \cos b \cos c \quad \ldots\ldots\ldots\ldots\ldots\ldots(1),$$
and also $\cos a = f(x', x'')/P'P''$. Then, from the equation
$$\Delta^2 f(x) = f(u')f(u'') - [f(u', u'')]^2,$$
by (iii), we have
$$\frac{\Delta P''}{UU'} = \frac{U''}{PP'} \Big/ \frac{U}{P'P''},$$
and hence
$$\sin C = \sin c/\sin a \ldots\ldots\ldots\ldots\ldots\ldots(2).$$

Also, from equations (ii), (iii), we have
$$f(x', x'') = \frac{f(u, u')f(u, u'')}{\Delta^2}, \quad f(x, x'') = -\frac{fu' \cdot f(u, u'')}{\Delta^2},$$
so that
$$-\frac{f(u, u')}{fu'} = \frac{f(x', x'')}{f(x, x'')},$$
which is
$$-\frac{f(u, u')}{UU'} = \frac{U'}{f(x, x'')} \Big/ \frac{U}{f(x', x'')},$$
namely
$$\cos C = \tan b/\tan a \ldots\ldots\ldots\ldots\ldots\ldots(3).$$

From the formulae (1), (2), (3), we deduce at once

$$\sin B = \sin b / \sin a, \quad \cos B = \tan c / \tan a, \quad \cos a = \cot B \cot C$$

together with

$$\cos B / \cos b = \sin C, \quad \cos C / \cos c = \sin B, \quad \tan B = \tan b / \sin c,$$
$$\tan C = \tan c / \sin b.$$

The sum of the angular intervals for a triangle. An immediate consequence of these formulae is that the angular intervals B, C have a sum which is greater than $\frac{1}{2}\pi$. For we have

$$\cos (B + C) = \cos B \cos C - \sin B \sin C = \sin B \sin C (\cos b \cos c - 1),$$

and $\cos b \cos c - 1$ is in all cases negative. Thus $\cos (B + C) < 0$, and therefore $B + C > \frac{1}{2}\pi$. Wherefore $A + B + C > \pi$.

And this last result is true for any triangle A, B, C, whether AB, AC be conjugate lines or not. For, let the lines drawn through A, B, C, each conjugate to the respectively opposite side, meet in the point P (Chap. I, Ex. 10). By our definition of the triangle, one at least of the lines AP, BP, CP contains a point of the opposite segment; for example, suppose that AP contains the point N of the segment BC. We can then apply the above argument to each of the triangles A, N, B and A, N, C. If the angular intervals BAN, NAC be called, respectively, ϕ and ψ, we have $\phi + B > \frac{1}{2}\pi$, $\psi + C > \frac{1}{2}\pi$, $\phi + \psi = A$, and hence $A + B + C > \pi$. This is an important result.

Formulae for a general triangle. Let the line through A which is conjugate to the line BC meet this in N, which is not necessarily a point of the segment BC. Denote by p the interval AN, and by ϕ, ψ, respectively, the angular intervals BAN, NAC, with the con-

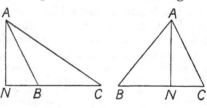

ventions introduced before in dealing with the right-angled triangle.

We have

$$\sin p = \sin c \sin B = \sin b \sin C ;$$

so that we have

$$\frac{\sin A}{\sin a} = \frac{\sin B}{\sin b} = \frac{\sin C}{\sin c} \quad \dots\dots\dots\dots\dots(1),$$

the first of these being introduced by a construction similar to that here described.

Further, using NC for $- CN$, and BN for $- NB$, with a similar convention for ϕ and ψ, when necessary, we have

$$\cos a = \cos (BN + NC) = \cos BN \cos NC - \sin BN \sin NC,$$
$$\cos b \cos c = \cos BN \cos NC \cos^2 p ;$$

hence

$$\cos a - \cos b \cos c = \cos BN \cos NC \sin^2 p - \sin BN \sin NC;$$

we have, however,

$$\sin b \sin c \cos A = \sin b \sin c \cos(\phi + \psi)$$
$$= \sin b \sin c (\cos \phi \cos \psi - \sin \phi \sin \psi),$$

while

$$\frac{\cos \phi}{\cos BN} = \sin B = \frac{\sin p}{\sin c}, \quad \frac{\cos \psi}{\cos NC} = \sin C = \frac{\sin p}{\sin b},$$

and

$$\sin \phi = \frac{\sin BN}{\sin c}, \quad \sin \psi = \frac{\sin NC}{\sin b};$$

hence

$$\sin b \sin c \cos A = \cos BN \cos NC \sin^2 p - \sin BN \sin NC.$$

Thus we infer

$$\cos a = \cos b \cos c + \sin b \sin c \cos A \dots\dots\dots\dots(2).$$

By a similar proof we have

$$\cos A = - \cos B \cos C + \sin B \sin C \cos a \quad\dots\dots\dots(3).$$

From these formulae it follows, also, that $\cos a - \cos(b + c)$, being equal to $2 \sin b \sin c \cos^2 \frac{1}{2}A$, is necessarily positive. Wherefore, if $\sigma = \frac{1}{2}(a + b + c)$, we see that each of the products $\sin \sigma \sin(\sigma - a)$, $\sin \sigma \sin(\sigma - b)$, $\sin \sigma \sin(\sigma - c)$ is positive; and from these we can deduce that $a + b + c < 2\pi$ and $b + c > a$ (Euclid, Book XI, Props. 21, 22). The latter of these two results is deducible from the former by considering the triangle whose angles are $A, \pi - B, \pi - C$; and the former is deducible from the previously obtained result, $A + B + C > \pi$, by considering the triangle whose vertices are the poles of the sides of the given triangle.

It should perhaps be remarked that though the formulae here obtained are identical with those usually given for the metrical trigonometry of the sphere, there is an important distinction between that case and this. For on the plane two points coincide if the interval between them is π, while two lines have only one point of intersection. What are two diametrically opposite points of the surface of a sphere are here replaced by a single point. The equivalent, in the present theory, of the geometry of the surface of a sphere, arises below (p. 196).

The case of a real conic as Absolute. When we take as absolute conic of reference a *real*, undegenerate, conic, of which we may suppose the equation to be $fx = 0$, where $fx = - x^2 - y^2 + z^2$, it is necessary, when we are discussing only the real points of the plane, to distinguish between points which are interior points of this conic, and those which are exterior. The former are characterised by the twofold property (above, p. 164) that every real line through an interior point meets the conic in two real points, while the tangents to the conic from the point are imaginary. For an exterior point, O, the tangents to the conic from O are real; some real lines drawn from O meet the conic in two real points, others in two imaginary points, and a line of the former sort is separated by the two real tangents from a line of the second sort.

It seems unnecessary to enter into great detail. We consider only real points which are interior to the conic; and our main object is to shew that the results obtained are those which were found by Lobatschewsky and Bolyai in their (metrical) so-called non-Euclidean geometry. Bibliographical references are given at the end of the volume (Note III).

Let the line joining two interior points, O, P, meet the conic in M and N, so chosen that N, O, P, M are in order; then the expressions of the symbols of O and P in terms of those of M and N will be of the forms $O = M + hN$, $P = M + kN$, where, if the point whose symbol is $M + N$ be an interior point of the conic, the real symbols h, k will be positive and $h > k$. Thus $(O, P; M, N)$, or $h^{-1}k$, is real and positive and less than 1, and we may take for $\log[(O, P; M, N)]$ a value -2σ, where σ is real and positive. Thus the interval θ as previously defined, by $(2i)^{-1}\log[(O, P; M, N)]$, is given by $\theta = i\sigma$. If, for fixed positions of N, O, M, the point P vary, from N, through O, to M, the value of σ varies from $-\infty$, through 0, to $+\infty$; and $\cos\theta$, $\sin\theta$ are, respectively, $\cosh\sigma$ and $i\sinh\sigma$. The conic, $-x^2 - y^2 + z^2 = 0$, meets the lines $x = 0$, $y = 0$ each in real points, but does not meet the line $z = 0$ in real points; thus the points $(1, 0, 0)$ and $(0, 1, 0)$ are exterior points, but the point $(0, 0, 1)$ is an interior point; for this point the expression $-x^2 - y^2 + z^2$ is positive; this expression is therefore constantly positive for all interior points. Similarly, if $u, = yz' - y'z$, $v, = zx' - z'x$, $w, = xy' - x'y$, be the co-ordinates of a line, the expression $-u^2 - v^2 + w^2$, which vanishes when the line touches the conic, has one sign for all lines which meet the conic in two real points, and the opposite sign for lines not meeting the conic in real points; the expression is negative for the line $x = 0$, and is thus negative for all lines which meet the conic in two real points. Again, if (x, y, z) and (x', y', z') be two interior points, and the signs of x, y, z, as of x', y', z', be properly taken, the expression $-xx' - yy' + zz'$ has a definite sign; for the polar of an interior point contains only exterior points; the expression is thus positive for any two interior points. We see thus that it is appropriate to replace the formulae

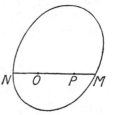

$$\cos\theta = (-xx' - yy' + zz')(fx \cdot fx')^{-\frac{1}{2}}, \quad \sin\theta = (fu)^{\frac{1}{2}}(fx \cdot fx')^{-\frac{1}{2}},$$

by the formulae

$$\cosh\sigma = (-xx' - yy' + zz')(fx \cdot fx')^{-\frac{1}{2}}, \quad \sinh\sigma = (-fu)^{\frac{1}{2}}(fx \cdot fx')^{-\frac{1}{2}},$$

replacing $(fu)^{\frac{1}{2}}$ by $i(-fu)^{\frac{1}{2}}$. All the square roots, as well as cosh σ and sinh σ, are now to be taken positively. These formulae can, if desired, be continued, to give the interval between an interior point, O, and an exterior point, P; the interval σ will then be of the form $\frac{1}{2}i\pi + \lambda$, where λ is real.

The interval between two lines which intersect at an interior point of the conic is to be taken with respect to the two imaginary tangents drawn to the conic from this point. Thus, as in the case of the imaginary conic considered above, the functions that enter are the trigonometrical functions, cos and sin, two lines which are at an interval π being coincident. In particular, as before, two conjugate lines are at an interval $\frac{1}{2}\pi$.

If we consider three points, A, B, C, all interior to the conic, such that AB and AC are conjugate lines, the polar line of A lies

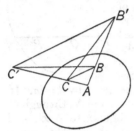

entirely exterior of the conic; and, if this meet AB and AC, respectively, in B' and C', these are the poles of AC and AB, respectively. The line, CB', through C, conjugate to CA, thus meets AB, at B', outside the conic and, therefore, the angular interval between CA and CB is less than $\frac{1}{2}\pi$. The same is true of the interval between BA and BC. Recalling, then, what was seen above, for the case when the absolute conic was imaginary, if we use a, b, c for the values of the interval σ belonging, respectively, to BC, CA, AB and B, C for the angle intervals ABC and ACB, the formulae for the triangle ABC are obtainable from those above given for an imaginary conic by merely replacing cos a, sin a, cos b, sin b, cos c, sin c by cosh a, sinh a, cosh b, sinh b, cosh c, sinh c, respectively. In particular,

$$\cos (B + C), = \sin B \sin C (\cosh b \cosh c - 1),$$

is necessarily positive. Thus we have $B + C < \frac{1}{2}\pi$, the angle between AB and AC being $\frac{1}{2}\pi$.

From this, as before, for any triangle A, B, C for which A, B, C are interior to the conic, we infer that $A + B + C < \pi$.

The angle interval, θ, between two lines $ux + vy + wz = 0$, $u'x + v'y + w'z = 0$, which intersect at a point (x, y, z), is such that sin θ is of the form $\Delta (fx)^{\frac{1}{2}}(fu . fu')^{-\frac{1}{2}}$; this interval is then zero for two lines which intersect on the absolute conic. It is thus possible, in this case, to have a triangle for which every one of the three angle intervals is zero, if we allow the vertices to be on the conic.

The sum of the angles of a triangle and a quadrangle of three right angles. If we recall (see above, Chap. IV, p. 158) that two pairs of real conjugate points, on the same line, separate one another, or do not separate one another, according as the line meets the absolute conic in imaginary, or in real, points, we can prove the important theorem for the sum of the angles of a triangle without the formulae we have developed; we give indications of how this may be done.

Let A, B, D, C be four points such that AB, CD both pass through the pole, A', of AC, while, also, AC, BD both pass through the pole of AB. The angles at A, B, C are then all equal to $\frac{1}{2}\pi$. Let B_1 be the pole of BD; this point, which is on BA', is separated from B by A and A' if AA' meet the absolute conic in imaginary points. In this case,

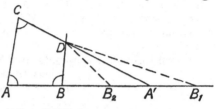

the angular interval between DB and DB_1 is $\frac{1}{2}\pi$, and, therefore, the angular interval between DB and DC, which is the fourth angle of the quadrangle $ABDC$, is greater than $\frac{1}{2}\pi$. When however the line AA' meets the absolute conic in real points, the pole of BD is not separated from B by A and A'; in this case, let this pole be denoted by B_2, instead of B_1. The angle BDB_2 being $\frac{1}{2}\pi$, the fourth angle, BDC, of the quadrangle $ABDC$, is now *less* than $\frac{1}{2}\pi$.

Bearing in mind the important result thus obtained for a quadrangle of which three angles are right angles, let A, B, C be any triangle with a right angle at A; and, for convenience, suppose the segments AB, AC are both less than $\frac{1}{2}\pi$. From B draw a line BN making the angle CBN equal to the angle ACB. Let O be the point of the segment CB such that the intervals CO and OB are equal; from O draw a line to the pole of AC, meeting this, at right angles, in M, and meeting BN in N. By comparing the

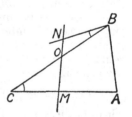

triangles CMO and BNO we infer, then, that ON is at right angles to NB. As the angles at A, M, N are right angles, we can by the preceding result, applied to the quadrangle $AMNB$, make an inference as to the angle at B, of this quadrangle; and this angle at B is equal to the sum of the angles ABC, ACB, of the triangle ABC.

Comparison with the Lobatschewsky-Bolyai geometry resumed. Now consider a line meeting the real absolute conic in the points M and N, and any point, A, interior to the conic, not lying on this line. We have seen that the interval, σ, between A and any real point of the conic is to be regarded as infinite, and that the interval between A and a point exterior to the conic, if considered at all, is to be regarded as imaginary. In accordance with our usual preconception of the position of a point as being determined by its distance from accessible points, it may seem natural to regard points which are exterior to the conic as not existent, or as at least inaccessible. If we do this, we shall regard all the lines through A which have intersections with the line MN as being limited by the lines AM and AN; and then we may speak of AM, AN as the *two* parallels to MN which can be drawn through A; their intersections with the line MN are at an infinite interval from A, and an infinite number of lines can be drawn through A which do not intersect MN (at any point interior to the conic). Let the line drawn through A conjugate to MN intersect this in H, a point

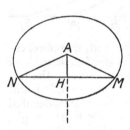

easily seen to be in the segment MN which is interior to the conic; let θ, ϕ denote, respectively, the angular intervals HAM and HAN, and let p be the value of the interval above denoted by σ, for AH. As the angles AMH, ANH are both zero, we infer, from the formulae

$$\sin \theta = \cos AMH \operatorname{sech} p,$$
$$\sin \phi = \cos ANH \operatorname{sech} p,$$

which we have obtained, that θ and ϕ are equal, both being $\sin^{-1}(\operatorname{sech} p)$. This would then be what, in the Lobatschewsky geometry, is called the *angle of parallelism*.

Further, if O be any fixed point interior to the absolute conic, and P a variable interior point such that the interval OP is constant, then P describes a conic having two contacts with the absolute conic, namely at the (imaginary) points where this conic is met by the polar line of O. This conic has the equation

$$(zz' - xx' - yy')^2 = (z^2 - x^2 - y^2)(z'^2 - x'^2 - y'^2) \cosh^2 \sigma,$$

where (x', y', z') are the coordinates of O, and σ is the constant interval OP. This conic is that called by Lobatschewsky a *circle*, of which (x', y', z') is the centre. If any chord, M', N', of this circle,

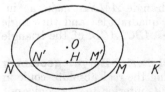

meet the absolute conic in M and N, and H, K be the two points which are the double points of the involution determined by the pairs M, N and M', N', the one of these points which is interior to the absolute conic being H, it is easy to see that the intervals $N'H$ and HM' are equal, and that the line drawn through H conjugate to $M'N'$ passes through O. In words, the lines, drawn

through the middle points of the chords of a circle at right angles
to these chords, all pass through the centre of the circle.

If through every point R, interior to the absolute conic, of a line
which meets the conic in two
real points, a line be drawn at
right angles to this line, and
thereon a point, P, be taken so
that the interval RP is always
the same (the point P being in
a sense to be explained, always
on the same side of the given
line) then the locus of P is
that called by Lobatschewsky
an *equidistant curve*. In fact, if
O, the pole of the given line, be

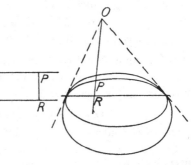

a point exterior to the given conic, the locus of a point P such that
the interval OP is constant, is precisely such an equidistant curve.

The corresponding locus, of P, may be considered when O is a
point of the absolute conic. This will then be a conic whose four
intersections with the
absolute conic all coin-
cide at one point. This
is the *horocycle* of Lo-
batschewsky. The lines
bisecting at right angles
the chords of the horo-
cycle are all parallel,
that is, they all pass
through a point (O) of

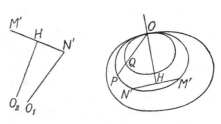

the absolute conic. This fact enables us to construct a horocycle
passing through a point N'; for draw through this a definite line
$N'O_1$; then draw any line $N'HM'$, the point H being such that the
angle $HN'O_1$ is the angle of parallelism for a line through N' which
is to be parallel to the line, HO_2, drawn through H at right angles
to $N'H$, namely $\cosh N'H$ is equal to $\operatorname{cosec}(O_1N'H)$ (see above).
The point M' is then such that $M'H = HN'$. If $N'O_1$ remain the
same, the angle $O_1N'H$ being allowed to vary, the locus of M' is a
horocycle. (This construction is given by Lobatschewsky, *Geometrische Untersuchungen zur Theorie der Parallellinien*, Reprint,
Berlin, 1887, § 31, p. 37.) If two horocycles have the same centre,
O, and a variable line through O meet these, respectively, in P and Q,
the interval PQ is constant; or, two concentric horocycles cut equal
segments on two parallel lines passing through their common centre
(Lobatschewsky, *loc. cit.* § 33, p. 39).

Case of degenerate conic as Absolute. In the case of a

conic represented in point coordinates by $z^2 + \kappa(x^2 + y^2) = 0$, or in line coordinates by $\kappa w^2 + u^2 + v^2 = 0$, say, respectively, $fx = 0$ and $Fu = 0$, the interval, p, between two points (x, y, z), (x', y', z') is, save for sign, such that

$$(fx \cdot fx')^{\frac{1}{2}} \cos p = zz' + \kappa(xx' + yy'),$$

$$(fx \cdot fx')^{\frac{1}{2}} \sin p = \{\kappa[\kappa(xy' - x'y)^2 + (xz' - x'z)^2 + (yz' - y'z)^2]\}^{\frac{1}{2}},$$

and the angle, θ, between two lines

$$ux + vy + wz = 0, \quad u'x + v'y + w'z = 0,$$

save for sign, is such that

$$(Fu \cdot Fu')^{\frac{1}{2}} \cos \theta = \kappa ww' + uu' + vv',$$

$$(Fu \cdot Fu')^{\frac{1}{2}} \sin \theta = \{\kappa(uw' - u'w)^2 + \kappa(vw' - v'w)^2 + (uv' - u'v)^2\}^{\frac{1}{2}}.$$

Consider what these become, supposing κ to be positive, when, first, κ approaches to zero, and, second, with the usual phraseology, κ becomes indefinitely great. In the former case the absolute conic becomes the point pair given by $z = 0$, $x \pm iy = 0$, or by $u^2 + v^2 = 0$. In this case the interval p becomes zero, but in such a way that $p \cdot \kappa^{-\frac{1}{2}}$ approaches to the limit, d, which is given by

$$d = \{(xz' - x'z)^2 + (yz' - y'z)^2\}^{\frac{1}{2}}/zz';$$

the interval, θ, however, remains finite, being, save for sign, such that

$$\cos \theta = (uu' + vv')/mm', \quad \sin \theta = (uv' - u'v)/mm',$$

where $m^2 = u^2 + v^2$, $m'^2 = u'^2 + v'^2$. When κ becomes infinitely great the absolute conic becomes the line pair given by $x \pm iy = 0$. The interval p becomes such that

$$\cos p = (xx' + yy')/nn', \quad \sin p = (xy' - x'y)/nn',$$

where $n^2 = x^2 + y^2$, $n'^2 = x'^2 + y'^2$; and the interval θ becomes zero, but in such a way that $\theta \cdot \kappa^{\frac{1}{2}}$ approaches to the limit ϕ given by

$$\phi = [(uw' - u'w)^2 + (vw' - v'w)^2]^{\frac{1}{2}}/ww'.$$

Consider the former case, obtained by supposing κ to vanish. If we regard the line $z = 0$ as consisting of those inaccessible points which we ordinarily speak of as being at infinity, this case may be held to include the familiar metrical geometry of the Greeks, on the basis of which measuring instruments at present in use are constructed. This usual metrical geometry evidently assumes that when two points are given, the segment which joins them is without ambiguity, being that segment of the line joining them which does not contain a point of the line $z = 0$. It will be natural enough to make that convention; for the degenerate conic of reference may be regarded as arising from an imaginary conic, $(\kappa > 0)$, and such conic contains no real point of any line considered; or may be regarded as arising from a real conic, $(\kappa < 0)$; and we have considered, in that case, only points lying within the conic. When that convention is made, only triangles being considered whose segments contain no point of the line $z = 0$, by putting $\alpha = \kappa^{\frac{1}{2}}a$, $\beta = \kappa^{\frac{1}{2}}b$, $\gamma = \kappa^{\frac{1}{2}}c$, we may deduce from the formulae, previously given (above, p. 178),

$$\sin \alpha'/\sin \alpha = \sin \beta'/\sin \beta, \quad \cos \alpha = \cos \beta \cos \gamma + \sin \beta \sin \gamma \cos \alpha',$$
$$\cos \alpha' = -\cos \beta' \cos \gamma' + \sin \beta' \sin \gamma' \cos \alpha,$$

the respective results

$$a^{-1}.\sin \alpha' = b^{-1}.\sin \beta', \quad a^2 = b^2 + c^2 - 2bc \cos \alpha', \quad \cos \alpha' = -\cos(\beta' + \gamma').$$

These may also be deduced independently. The last of these follows from $\alpha' + \beta' + \gamma' = \pi$, which can be proved, as by Euclid, from the

fact that two lines BA, CA, which meet $z = 0$ in the same point A, are at equal angular intervals, ABT, ACT, with a line BC meeting them in B and C, respectively. As before, this assumes that the seg-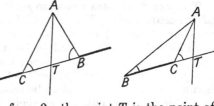

ment BC contains no point of $z = 0$; the point T is the point of the line $z = 0$, on the line BC, on the *other* segment BC.

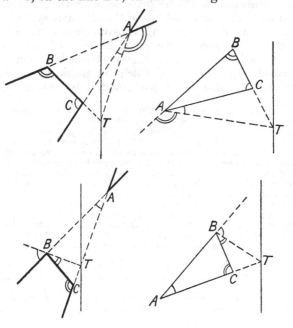

It follows from this that if we consider an (improper) triangle of which the segments AB, AC contain points of $z = 0$, but the segment BC contains no such point, and A, B, C denote the angular intervals BAC, CBA, ACB, then $A + \pi - B + \pi - C = \pi$, or $B + C = \pi + A$. In particular when $A = \frac{1}{2}\pi$, that is, when the points of intersection of AB and AC with $z = 0$ are harmonic in regard to the absolute points, I, J, lying on $z = 0$, we have $A + B + C = 2\pi$.

It appears sufficiently clear from this that any statement of the ordinary Euclidean geometry, even one which has reference to measurement, continues to hold when the so-called line at infinity is replaced by any line of the plane, and the so-called circular points at infinity are replaced by any two conjugate imaginary points lying thereon[1].

The arbitrariness and comparison of the several methods of measurement. When intervals, between points and between lines, are measured by means of an imaginary conic, the figure considered is sometimes said to lie in an *elliptic* plane; when the conic is real, in an *hyperbolic* plane; and, when the conic of reference is a point pair, in a *parabolic*, or Euclidean plane. The phraseology is convenient; but it is apt to be misconceived. *The plane and the geometry* (though not the symbols, or numbers, associated with intervals) *are the same in all the three cases.* Statements in regard to the intervals are to be considered as abbreviated statements in regard to particular combinations of the algebraic symbols merely; and we have sought to shew that the symbolism is not a necessary, though it is sometimes a convenient, support of the geometrical theory. No deduction of a really geometrical kind can be legitimately based on statements of which any particular conception of distance forms an indispensable part; such statements are equivalent

---

[1] The proof given in Volume I, pp. 120, 121, that Desargues' theorem does not follow from Plane propositions of incidence only, was originally given, by Moulton, for the case of the Euclidean plane, that is, under the hypothesis that all the lines considered lie 'on one side of' the line YU; and it is with that tacit hypothesis that the proof is reproduced, though then two lines have not necessarily a point of intersection. The statement, in the ninth and tenth lines from the bottom of page 120, that the line HQ is to be continued beyond Q is an error, kindly pointed out to the writer by Dr W. Burnside; only the line HP is to be indefinitely continued, beyond P. But we may cease to observe the hypothesis referred to. Then any of the broken lines occurring in the proof will have *two* points upon the line YU; in particular the line HP, continued, will reappear 'on the other side of' YU, and meet this in M. If it were allowed to the line HQ to be continued beyond Q, it might well contain a point R such that the segment PR did not contain a point of the segment YU; in the theory, as it is meant, such a point R does not lie on the broken line $QHPM$.

It is of course assumed in this theory that if Desargues' theorem were a consequence of the Plane propositions of incidence only, it would be impossible, by the addition of any further restrictive hypothesis not contradictory to the propositions of incidence, such as those here introduced, to find a case in which Desargues' theorem does not continue to hold. Moulton's theory appears to find such a case.

only to statements in regard to the behaviour of particular measuring instruments, which must rest on physical hypotheses.

In illustration of this general remark, we may, if we please, apply two of the systems of measurement simultaneously. Thus, let A, B, C be any three points, and consider a triangle formed with these, in the sense explained above when discussing the reference to an imaginary conic. Let the segment intervals, BC, CA, AB, with reference to a certain imaginary absolute conic, be denoted, respectively, by α, β, γ, and the angle intervals, BAC, CAB, ABC, corresponding thereto, be denoted, respectively, by α', β', γ'. If we take any three points as reference points for coordinates, and two points I, J which, relatively to these reference points, are given by $z = 0$, $x \pm iy = 0$, we can also measure the segments and angles, as explained above, with reference to the point pair I, J; let them be denoted then, respectively, by a, b, c, A, B, C. Then *both* the equations

$$\cos \alpha = \cos \beta \cos \gamma + \sin \beta \sin \gamma \cos \alpha',$$

and
$$a^2 = b^2 + c^2 - 2bc \cos A,$$

are true.

An interesting application of this remark, which has some importance of its own, may be made. This expresses the interval, measured with reference to the imaginary conic $x^2 + y^2 + z^2 = 0$, between two lines, by means of an integral which is to be evaluated with measurements of distance and angle that are made relatively to the point pair, I, J, given by $z = 0$, $x \pm iy = 0$. Represent x/z, y/z, respectively, by X and Y; an element of area, with the Euclidean reference, is, we may assume, on the basis of familiar notions, given by $dX\,dY$, or by $r\,dr\,d\theta$, where r is the Euclidean distance from a point, and θ the Euclidean angle from a fixed line. If two lines, intersecting in O, be given, of equations $ux + vy + wz = 0$, $u'x + v'y + w'z = 0$, these may be regarded as defining a region, (P) and (P'), over which the integral

$$\iint \frac{dX\,dY}{(1 + X^2 + Y^2)^{\frac{3}{2}}}$$

may be taken (every element $dX\,dY$ being positive). It will be found that the *integral is* then *equal to twice the interval between the lines*, measured in regard to the conic $x^2 + y^2 + z^2 = 0$. Of this statement we may give indications of a straightforward proof. Let d denote the Euclidean distance from the origin z, or $(0, 0, 1)$, to the intersection, O, of the lines, r denote the Euclidean distance from O to a point, P, of the region of integration, and θ the Euclidean angle between the line zO, and the line OP. The integral is then

$$\int_{\phi}^{\phi'} d\theta \int_{0}^{\infty} r dr \{(1 + d^2 + r^2 + 2dr \cos \theta)^{-\frac{3}{2}} + (1 + d^2 + r^2 - 2dr \cos \theta)^{-\frac{3}{2}}\},$$

where ϕ is the angle between the line zO and one of the given lines, ϕ' the angle between zO and the other given line. This form of the integral is obtained by pairing together points, P and P', on a line through O, in regard to which the harmonic conjugate of O is on the line $z = 0$; the factor $r \cos \theta$, arising for P, corresponds to a factor $r \cos (\theta + \pi)$, or $(-r) \cos \theta$, arising for P'. To carry out the integration, we may notice that

$$\int x dx [(x - a)^2 + b^2]^{-\frac{3}{2}} = [ax - (a^2 + b^2)] b^{-2} [(x - a)^2 + b^2]^{-\frac{1}{2}},$$

and utilise the identity

$$\cos^{-1} \frac{u'V - v'U}{s'\Delta} - \cos^{-1} \frac{uV - vU}{s\Delta} = \cos^{-1} \frac{uu' + vv' + ww'}{ss'},$$

where

$$U = vw' - v'w, \quad V = wu' - w'u, \quad s^2 = u^2 + v^2 + w^2,$$
$$s'^2 = u'^2 + v'^2 + w'^2, \qquad \Delta^2 = U^2 + V^2.$$

The integral will then be found to become

$$2 \cos^{-1} [(uu' + vv' + ww')/ss'].$$

On the basis of this result, we may speak of the *extent*, relatively to the imaginary conic, of a region of the plane, meaning thereby the value of the integral, evaluated, with the parabolic measurements, as here, over this region. In this sense, any two lines enclose a finite extent, equal to $2A$, where A is the interval between the lines, measured by the imaginary conic. The two lines enclose two regions, which are complementary; the sum of the extents of these two regions is that of the whole plane, which is thus finite. The angular interval, in regard to the imaginary conic, for the other region, is $\pi - A$, and the extent of this is $2\pi - 2A$. Thus 2π is the extent of the whole plane.

Thus also the extent of a triangle, in terms of the angular

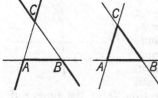

intervals of it, with reference to the imaginary conic, is $A + B + C - \pi$. For, if the extent of the triangle be denoted by Δ, the whole plane is made up of four regions whose respective extents are $2A - \Delta$, $2B - \Delta$, $2C - \Delta$, and Δ; putting the sum of these equal to 2π, we find

$$\Delta = A + B + C - \pi.$$

Conversely, we may avoid the use of the integral, by defining the extent of a triangle as being $A + B + C - \pi$, allowing this to apply also to the triangle whose angular intervals are $C, \pi - A, \pi - B$, and assuming that the extent of a composite region is the sum of the extents of its component regions. For the extent between the lines CA, CB we should thus obtain

$$[A + B + C - \pi] + [C + \pi - A + \pi - B - \pi], \text{ or } 2C.$$

Ex. 1. Shew that the extent of a *circle*, relatively to the imaginary conic, that is of the conic whose equation is

$$(xx' + yy' + zz')^2 = (x^2 + y^2 + z^2)(x'^2 + y'^2 + z'^2) \cos^2 r,$$

is $2\pi (1 - \cos r)$.

Ex. 2. If σ be the interval, relatively to the imaginary conic, between two points, prove that the integral $\int d\sigma$, extended along the circle, has the value $2\pi \sin r$. (Cf. p. 200.)

The remarks in the present section are largely anticipatory; it will be understood that they are intended to be suggestive, and are not complete.

Plane metrical geometry by projection from a quadric surface. Spherical geometry. In Chap. III we have considered the theory of the circle, or of geometry with two absolute points of reference. In the present chapter we have considered the theory with reference to an absolute conic. These are in pursuance of the plan we have adopted, of devoting the present volume to geometry in one plane. We shall see in detail, in a subsequent volume, that a plane in which two absolute points are given is more properly to be regarded as the projection of a surface, called a quadric, existing in space of three dimensions, which has the property of being met in two points by an arbitrary line. The centre of projection lies on this surface. Similarly, a plane in which an absolute conic is given is naturally suggested by the projection of the points of a quadric, from a centre of projection not lying on the quadric. It appears, therefore, to be necessary to make some remarks here in regard to a quadric surface; as we are to deal with geometry in three dimensions in the next volume, these remarks will be brief, and incomplete.

We have already, in Volume I, p. 82, constructed, from four points A, B, C, D which do not lie in a plane, a point whose symbol, in terms of the symbols of these points, is

$$P = \lambda A + \mu B + \lambda\mu C + D;$$

if we denote this by $P = xA + yB + zC + tD$, we have $xy = zt$; conversely, any point whose *coordinates*, x, y, z, t, relatively to A, B, C, D, satisfy this equation, is such a point as P, for suitable values of λ and μ. It is the aggregate, or locus, of all such points which we speak of here as a quadric surface. For our purpose it will be convenient to use ξ, η instead of x, y, subject to $x = \xi + i\eta$, $y = \xi - i\eta$, so that the equation takes the form $\xi^2 + \eta^2 = zt$. We can shew that the surface contains an infinite number of lines, which break up into two systems, there being one line of each system through every point of the surface. Every line of either system, in fact, meets every line of the other system, but two lines of the same system do not intersect. For, first, whatever θ may be, every point (ξ, η, z, t) which is such that $\xi + i\eta = \theta t$, $\xi - i\eta = \theta^{-1}z$, is evidently such that $\xi^2 + \eta^2 = zt$, and is, therefore, a point of the surface; such a point has the symbol $\theta t A + \theta^{-1}z B + z C + t D$, and is a point of the line joining the points whose symbols are $\theta^{-1}B + C$, $\theta A + D$; the two equations taken together thus represent this line. By taking different values of θ, we thence obtain a system of lines. If $(\xi_0, \eta_0, z_0, t_0)$ be any point of the surface, the equal values, $(\xi_0 + i\eta_0) t_0^{-1}$, $z_0 (\xi_0 - i\eta_0)^{-1}$, give a value of θ for which the line contains this point. It is easy, moreover, to verify that two lines of this system, for the respective values θ, ϕ of the parameter, given by $\xi + i\eta = \theta t$, $\xi - i\eta = \theta^{-1}z$ and $\xi + i\eta = \phi t$, $\xi - i\eta = \phi^{-1}z$, do not meet.

Similarly, the surface contains another system of lines, of which one is given by $\xi - i\eta = \phi t$, $\xi + i\eta = \phi^{-1}z$, one of which passes through every point of the surface. No two of these intersect. But this line meets that, of the former system, given by $\xi + i\eta = \theta t$, $\xi - i\eta = \theta^{-1}z$, in the point given by $\xi + i\eta = \theta$, $\xi - i\eta = \phi$, $z = \theta\phi$, $t = 1$.

It is easy to prove that any line of the threefold space contains two points of the quadric, unless it consist wholly of points belonging thereto; for the substitution in $xy = zt$ of values $x_1 + \sigma x_2$, $y_1 + \sigma y_2$, $z_1 + \sigma z_2$, $t_1 + \sigma t_2$, respectively for x, y, z, t, leads to a quadric equation for σ. Further, that the points of an arbitrary plane, which lie on the quadric, are the points of a conic; for, let any line of this plane meet the quadric in the two points P and R, of coordinates (x_1, y_1, z_1, t_1) and (x_3, y_3, z_3, t_3); and let Q (x_2, y_2, z_2, t_2) be another point of this plane, not on the line PR: if, then, in $xy = zt$, we replace x, y, z, t, respectively, by $x_1 + \theta x_2 + \phi x_3$, and similar expressions, we find that ϕ is expressible in terms of θ; by choosing Q suitably, this leads to a symbol, for any point of the quadric lying in the plane, of the form $P + m\theta Q + n\theta^2 R$. This shews that all such points lie on a conic. In particular, the plane containing the two lines of the quadric which meet in any point, P, of the surface, contains no other points of the quadric than the points of these two lines. This plane is called the tangent plane of the quadric at P. The lines are called the generating lines, or generators, of the surface, at P.

Any equation which is linear (and homogeneous) in the coordinates, ξ, η, z, t, of a point, expresses that this point lies on a definite plane; this equation may then be called the equation of the plane. If the equation be $-ct + 2f\eta + 2g\xi + hz = 0$, the point $(\xi + i\eta)A + (\xi - i\eta)B + zC + tD$, being, in virtue of this equation, the same as

$$c(\xi + i\eta)A + c(\xi - i\eta)B + czC + (2f\eta + 2g\xi + hz)D,$$

or $\quad \xi(cA + cB + 2gD) + \eta(icA - icB + 2fD) + z(cC + hD),$

is a point of the plane containing the three points whose symbols are, respectively,

$$c(A + B) + 2gD, \quad ic(A - B) + 2fD, \quad cC + hD;$$

by proper choice of c, f, g, h, this is any plane whatever.

In particular, the three equations $\xi = 0$, $\eta = 0$, $z = 0$ represent, respectively, three planes. And these intersect in the point of co-ordinates $(0, 0, 0, 1)$, that is, the point D. This is a point of the quadric, since these coordinates satisfy the equation $\xi^2 + \eta^2 = zt$. The line which joins this point, D, to any point, of coordinates (ξ, η, z, t), meets the plane whose equation is $t = 0$, in the point $(\xi, \eta, z, 0)$. This point is the projection, from D, of the point (ξ, η, z, t). Any point of the quadric, other than D itself, can thus be made to correspond, by projection, to a point of the plane $t = 0$. But the line which is expressed by the two equations $\xi + i\eta = 0$, $z = 0$, is wholly on the quadric; and this line passes through D; every point of the quadric lying on the line is, therefore, projected from D into the same point of the plane $t = 0$. So, likewise, is every point of the quadric lying on the line expressed by the two equations $\xi - i\eta = 0$, $z = 0$. Every point of the plane $t = 0$, other than the two points so obtained, is the projection of a definite point of the quadric. These two exceptional points of the plane $t = 0$ are, necessarily, themselves points of the quadric; if we denote them by I and J, we may regard the point D of the quadric as projecting into the line IJ.

In this projection, the points of the quadric which lie on any plane passing through D project into the points of a line, in the plane $t = 0$, and conversely. A plane not passing through D is intersected by the generators of the surface at D (whose equations are $\xi + i\eta = 0$, $z = 0$ and $\xi - i\eta = 0$, $z = 0$) in two points, say, I' and J'; the points of the quadric lying on such a plane project, therefore, into the points of a curve, in the plane $t = 0$, which passes through the points I and J. If the plane in question be that given by the equation $-ct + 2f\eta + 2g\xi + hz = 0$, we see, elimi-nating t by means of $\xi^2 + \eta^2 = zt$, that the equation of this curve is $-c(\xi^2 + \eta^2) + 2f\eta z + 2g\xi z + hz^2 = 0$. The curve is that we have, in Chap. III, called a circle, relatively to the points I, J. In particular, when the plane contains the point D, we have $c = 0$, and the circle degenerates into the line, in the plane $t = 0$, given by $z = 0$, together with another line.

Thus the geometry in a plane, relative to two absolute points, which we have considered in Chap. III, may be regarded as the geometry dealing with the points of a quadric surface. A circle in the plane is thereby replaced by a plane section of the quadric surface.

Consider, from this point of view, the measurement of angle between two lines in the plane $t = 0$, relatively to the points I, J. Let the lines intersect in Q, and meet the line IJ, respectively, in the points M, N. The measurement, we have explained (above, p. 167), is by means of the range M, N, I, J. The four lines QM, QN, QI, QJ, of the plane $t = 0$, are projected from D by means of four planes which intersect in the line DQ. Any plane, not containing the line DQ, is met by these four planes in a flat pencil of four lines which is related to the pencil $Q(M, N, I, J)$, in the sense explained in Vol. I; in particular we may take the tangent plane of the quadric at D, which is met by these four planes in the pencil $D(M, N, I, J)$. Let DQ meet the quadric surface in Q' (beside D); there is, at Q', one generator which meets the generator DI, say in I', and another generator which meets the generator DJ, say

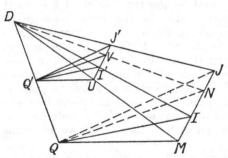

in J'. It is easy to verify that the tangent line of any plane section of the quadric, at any point, lies in the tangent plane of the surface at that point. The planes DQM, DQN give rise to two conics lying on the quadric surface, which have the points Q' and D for common points; the tangent at Q' of the former conic meets the tangent of this at D, in a point, say, U, which is on the line DM in which the plane DQM meets the tangent plane, DIJ, at D; similarly, the tangent at Q' of the latter conic meets the tangent of this at D, in a point V lying on DN. These points U, V are in the tangent plane of the quadric at Q'. Thus the pencil $Q'(U, V, I', J')$ is related to the pencil $Q(M, N, I, J)$. Two tangent planes can be drawn to the quadric surface from an arbitrary line, as we may see by taking the two points where the line meets the quadric, and considering the intersections of the generators of the surface at one of these points with the generators of the surface at the other of these two points. In our figure, as $I'D, I'Q'$ are generators, and $J'D, J'Q'$ are generators, the planes $DI'Q'$ and $DJ'Q'$ are the tangent planes of the surface which can be drawn through the line DQ; thus the axial pencil of planes which has been spoken of, consists of the two planes through DQ containing the lines QM and QN, together with the two tangent planes to the surface from the line DQ. It is by considering the two former planes, with respect to the two latter, that the angle between the lines QM, QN is measured.

It can be shewn that if, through any point E, a line be drawn to meet the quadric surface in R and S, and the point F be taken thereon harmonic to E in regard to R and S, then the locus of F, for all such lines drawn through E, is a plane. This plane, say e, is called the polar plane of E in regard to the quadric. For the quadric $\xi^2 + \eta^2 = zt$, its equation is $2(\xi\xi' + \eta\eta') = zt' + z't$, where (ξ', η', z', t') are the coordinates of E. If the polar plane of one point, E, contain a point, F, then the polar plane of F contains the point E; two such points, E, F, are said to be conjugate in regard to the quadric. To any arbitrary plane there is one point, called its pole, of which the plane is the polar plane; if the pole of one plane lie on another plane, the pole of this other lies on the former; two such planes are said to be conjugate. To any plane drawn through a given line, there corresponds another plane passing through the line, conjugate to the former plane; the pairs of planes so obtainable form an axial pencil in involution, of which the double elements are the two tangent planes to the quadric which can be drawn through the given line.

With this, we can state in another way the condition adopted in Chap. III for the lines QM, QN to be at right angles to one another in regard to I and J, which was that M and N should be harmonic conjugates in regard to I and J. It requires that the planes DUQ', DVQ' should be harmonic in regard to the tangent planes, $DI'Q'$, $DJ'Q'$, drawn to the quadric from the line DQ. Thus, from what is said above, the condition is that the planes DQM and DQN should be conjugate to one another in regard to the quadric.

Now suppose we take two circles in the plane $t = 0$, intersecting in the point Q, and in another point, R, in addition to I and J. We may consider the pencil of four lines, consisting of QI, QJ and the two tangents of these circles at Q. Let DQ and DR meet the quadric again, beside D, in Q' and R', respectively. The pencil of four lines, spoken of, passing through Q, is, we have seen, related to a pencil of four lines through Q', lying in the tangent plane of the quadric at Q'; these lines are the generators of the quadric at Q' (corresponding to QI, QJ), and the tangents at Q' of the two plane sections of the quadric which, as we have seen, project, from D, into the two given circles of the plane $t = 0$. Through these four lines, in the tangent plane at Q', we can draw four planes all passing through the line $Q'R'$, so determining an axial pencil of planes related to the plane pencil in question. Of these planes, one will be a plane containing one of the generators at Q' and, also, that generator at R' which intersects this ; this plane will be a tangent plane to the quadric drawn through the line $Q'R'$; another of the four planes will be that containing the other generator at Q', of the quadric, together with the generator at R' which intersects this, this being the other tangent plane to the quadric drawn from the line $Q'R'$. The other two planes of the axial pencil through $Q'R'$ are the planes of the two sections of the quadric which project, from D, into the two given circles. In accordance with what is said at the beginning of this chapter, the angle between the tangents at Q, of the two given circles, may be measured by the plane pencil, of centre Q, by taking these tangents with respect to QI and QJ. By what we have now said, it may also be measured by taking the plane sections of the quadric, through the line $Q'R'$, which project from D into the given circles, and considering these planes with respect to the two tangent planes of the quadric which can be drawn from the line $Q'R'$. Incidentally, it appears that the circles cut at the same angle at both their common points, Q and R. In particular, the circles cut at right angles it the two plane sections of the quadric, which project from D into these circles, are in planes which are conjugate to one another in regard to the quadric.

In these discussions, the lines and circles whose inclinations have been considered lie in the plane $t = 0$, which is in fact a tangent plane of the quadric, namely at the point $(0, 0, 1, 0)$. It is true, however, that a plane section of the quadric projects, from a point D of the quadric, into a circle upon any plane whatever, if the two absolute points for this plane are taken to be the points in which the generators of the quadric, at D, meet this plane.

Let us now replace the coordinates z and t by coordinates ζ, τ by means of $z = \tau + \zeta$, $t = \tau - \zeta$; the equation of the quadric then takes the form $\xi^2 + \eta^2 + \zeta^2 = \tau^2$. There is in the plane $\tau = 0$ a conic, lying on the quadric, given by $\xi^2 + \eta^2 + \zeta^2 = 0$. This conic meets an arbitrary plane, ϖ, of the threefold space in two points. Let us agree to take for the two absolute points of this plane, ϖ, in the sense discussed at length in Chap. III, these two points of this quadric. For a tangent plane of the quadric, say, at Q', these two points, lying on the quadric and on this plane, are the two points in which the plane $\tau = 0$ is met by the generators of the quadric at Q'. The effect is, then, that the angle between the tangent lines of the two plane sections of the quadric, at a point Q' where these sections intersect, as we have been led to consider it, is in fact obtainable by the rule, applied to the tangent plane of the quadric at Q', which we originally (p. 167) adopted for geometry in a plane ; but, at the same time, the suggestion clearly arises of considering the interval between two points, of the plane $\tau = 0$, by a reference to a fixed conic lying in this plane.

Metrical geometry in regard to an absolute conic. We proceed now to consider this measurement in regard to a fixed conic from the point of view we have reached by introducing the quadric surface.

Denote the point $\xi = 0$, $\eta = 0$, $\zeta = 0$, which does not lie on the quadric, whose equation is $\xi^2 + \eta^2 + \zeta^2 = \tau^2$, by O. It is easily seen that the plane $\tau = 0$ is the polar plane of O in regard to the quadric. Any point, say, Q', of the quadric, may be projected from O on to the plane $\tau = 0$, giving a point, say, Q_1. All points, Q', of the quadric which lie on a plane passing through O, give rise, then, to points, Q_1, lying on a line of the plane $\tau = 0$. Consider two plane sections of the quadric whose planes pass through O; the common points, say Q' and R', of these two sections of the quadric, will then lie in line with O. We have seen that the angle between the tangent lines, at Q', of these sections, obtained by considering these lines with respect to the generators of the quadric at Q', may be obtained also by considering the planes of the two sections with respect to the two tangent planes of the quadric which can be drawn through the line $Q'R'$. Now consider the four lines in which these planes meet the plane $\tau = 0$; as $OQ'R'Q_1$ are in line, these four lines meet in the point Q_1 of the plane $\tau = 0$. Two of these lines, corresponding to the tangent planes of the quadric drawn through $Q'R'$, are the intersections with $\tau = 0$ of the planes joining O to the generators of the quadric at Q'; the other two of the four lines are the intersections with $\tau = 0$ of the planes of the two given sections. We can shew that the two former lines are the tangent lines, from Q_1, to the conic of the plane $\tau = 0$ given by $\xi^2 + \eta^2 + \zeta^2 = 0$. For let one of the generators at Q', of the quadric, meet the plane $\tau = 0$, say, in H; this point is on the section of the quadric by the plane $\tau = 0$, that is, on the conic $\xi^2 + \eta^2 + \zeta^2 = 0$. The plane $\tau = 0$ being the polar plane of O, it is easy to see that the tangent plane of the quadric at H passes through O; this tangent plane contains the generator $Q'H$. Thus the plane joining O to this generator meets $\tau = 0$ in a line which touches the conic $\xi^2 + \eta^2 + \zeta^2 = 0$ at H, as was said. We are thus led from the measurement of the angle, at Q', between two lines, meeting at this point, lying in the tangent plane at Q', by means of two absolute points of this plane, to the measurement of the angle, at Q_1, between two lines, meeting at this point, lying in the plane $\tau = 0$, by means of the tangent lines drawn from this point to the conic $\xi^2 + \eta^2 + \zeta^2 = 0$.

Next consider any two points of the quadric, say, Q' and K'. Let them be projected from O into the two points, Q_1 and K_1, of the plane $\tau = 0$. Suppose that we measure the interval between the points Q' and K', as points of the quadric surface, by the angle between the lines OQ' and OK', taken in respect to the two absolute points of the plane $Q'OK'$, these being, as before, the points where this plane meets the conic $\tau = 0$, $\xi^2 + \eta^2 + \zeta^2 = 0$. We are thus led to measure the interval between the points, Q_1 and K_1, of the plane $\tau = 0$, by considering these with respect to the two points in which their joining line meets the absolute conic of their plane.

Thus, finally, the plane metrical geometry we have explained in this chapter, relatively to a fixed absolute conic, which gives an interval between two lines, and an interval between two points, is seen to be obtainable by projection of a geometry upon the quadric surface, the centre of projection, O, not being on the quadric. Lines of the plane are replaced by sections of the quadric, whose planes pass through the centre of projection, O. The interval between two points on the quadric surface is understood to be the interval between the lines joining these to O, taken with respect to the absolute points of the plane containing these joining lines; the interval between two lines of the plane is replaced by the interval between two lines lying in a tangent plane of the quadric surface, again taken with respect to the absolute points of this plane.

But this reduction of the metrical plane geometry, with respect to an absolute conic, to metrical geometry on a quadric surface, with respect, in the case of every plane concerned, to two absolute points in that plane, calls for a further important remark. While any point, Q', of the quadric, gives rise, by projection from O, to a single point, Q_1, of the plane, this last gives rise, conversely, not only to Q', but also to the second point, R', in which the line OQ_1 meets the quadric. Thus, whereas two points of the plane coincide when the interval between them, in regard to the absolute conic, is π (or any odd multiple of π), the corresponding points of the quadric surface do not then coincide, but are two points in line with O. Two points of the quadric surface coincide only when the interval between them is 0, or 2π, or an even multiple of π. The geometry of the plane, as developed in this book, is in fact that of lines, extended in both directions, passing through a point outside the plane. In the geometry on the quadric surface, which we may call *spherical* geometry, any such line gives rise to *two* points. In the plane two lines have one point in common; on the quadric surface, two plane sections are curves having *two* points in common. With proper conventions, the metrical formulae we have found for a triangle, when the measurements are with respect to an imaginary conic, continue to hold on the quadric given by the equation $\xi^2 + \eta^2 + \zeta^2 = \tau^2$, in which the coordinates ξ, η, ζ, τ are real; the arcs between two points being taken on planes through the point $(0, 0, 0, 1)$, and measured, as explained, with respect to the two points in which the planes meet the conic $\xi^2 + \eta^2 + \zeta^2 = 0$, $\tau = 0$; while the angles between two such arcs are measured, on the tangent plane of the quadric, at the point of intersection of the arcs, with respect to the two points in which this plane meets the same conic.

But there are many cases in which the distinction between the plane geometry and the spherical geometry is of importance. As a single example we may refer to the following: On the quadric $\xi^2 + \eta^2 + \zeta^2 = \tau^2$, consider the four points, A, B, C, D, of coordinates, respectively, $(-a, b, c, 1)$, $(a, -b, c, 1)$, $(a, b, -c, 1)$, $(-a, -b, -c, 1)$, where a, b, c are arbitrary real symbols such that $a^2 + b^2 + c^2 = 1$. These can be joined in pairs by six arcs, lying on the quadric in planes which pass through the point $(0, 0, 0, 1)$, forming four triangles ABC, BCD, CAD, ABD, in such a way that if the angles of the first of these be called A, B, C, those of the others are, respectively, C, B, A; A, C, B and B, A, C; the four triangles, which together exhaust the whole surface of the quadric, are in fact equal to one another, in angles and in arcs. The three angles A, B, C are such that $A + B + C = 2\pi$. When, however, this figure is projected from O, the point $(0, 0, 0, 1)$, on to the plane $\tau = 0$, while we get, from A, B, C, the triangle A_1, B_1, C_1, for which the angles are such that $A_1 + B_1 + C_1 = 2\pi$, the point D_1, arising from D, is an interior point of this, namely is such that every line through it meets two of the lines $B_1 C_1$, $C_1 A_1$, $A_1 B_1$. (See below, p. 207.)

Representation of the original plane upon another plane. In the preceding section we have passed from a point, Q, of the plane, by projection from a point D, lying on a quadric surface, to a point, Q', of this surface. And then, by projection from a point, O, not lying on the quadric, to another point Q_1, lying on another plane (that denoted by $\tau = 0$). If, with O and D given, we consider the reverse process of passing from Q_1 to Q, the point Q_1 will give rise, by projection from O, to two points of the quadric surface, say, Q' and R', and each of these will give, by projection from D, a single point of the original plane; so that we get two points, Q and R, corresponding to Q_1. But the results of the process will be found to be interesting, nevertheless. In particular, it will appear that the *distance* used by Riemann in his celebrated Dissertation of 1854, *Ueber die Hypothesen welche der Geometrie zu Grunde liegen* (Werke, 1876, p. 254), which at first sight appears to require that a system of metrical coordinates has been established beforehand, is in fact identical with the interval with respect to a conic which we have considered in the present chapter. Further, the origin of one particular representation of Lobatschewsky's Geometry (used, for example, by Poincaré, in his remarkable papers on Automorphic Functions, *Acta Math.* I, 1882, p. 7) will be made clear.

The absolute conic being supposed to be given by one of the two equations $z^2 + x^2 + y^2 = 0$, $z^2 - x^2 - y^2 = 0$, and, in the latter case, limiting ourselves to real values of x, y, z for which $z^2 - x^2 - y^2$ is positive, as in the earlier part of this chapter, let us associate, with each point, (x, y, z), of the plane the positive square root, $(z^2 + x^2 + y^2)^{\frac{1}{2}}$, or $(z^2 - x^2 - y^2)^{\frac{1}{2}}$. If we denote this square root by t, this comes to considering the point (x, y, z, t), of the quadric surface expressed, respectively, in the two cases, by the first or second of the equations $x^2 + y^2 + z^2 = t^2$, $z^2 - x^2 - y^2 = t^2$. This point, (x, y, z, t), is one of the two points in which the quadric is met by the line joining the point (x, y, z), of the plane, to the point $(0, 0, 0, 1)$. For the former quadric surface, this point is an interior point, in the sense that every line through it meets the quadric in two real points. For the latter quadric, the point $(0, 0, 0, 1)$ is an exterior point, only those lines through it which contain points for which $z^2 - x^2 - y^2 > 0$ meeting it in real points.

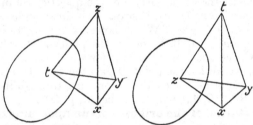

In the case of points of the quadric $x^2 + y^2 + z^2 = t^2$, if we define ξ, η, respectively, by $\xi = 2x/(z + t)$, $\eta = 2y/(z + t)$, we may take, for the homogeneous coordinates, x, y, z, t, of a point of the quadric, respectively,

$$x = \xi, \quad y = \eta, \quad z = 1 - \tfrac{1}{4}(\xi^2 + \eta^2), \quad t = 1 + \tfrac{1}{4}(\xi^2 + \eta^2);$$

if we take $\lambda = \tfrac{1}{2}(\xi + i\eta)$, $\mu = \tfrac{1}{2}(\xi - i\eta)$, these are the same as

$$x = \lambda + \mu, \quad y = -i(\lambda - \mu), \quad z = 1 - \lambda\mu, \quad t = 1 + \lambda\mu.$$

Noticing that

$$\xi = x + t.0, \quad \eta = y + t.0, \quad 2 = z + t.1, \quad 0 = t + t(-1),$$

we may, if we wish, regard $(\xi, \eta, 2, 0)$ as the coordinates of a point, in the original plane of x, y, z, obtained by projecting the point, (x, y, z, t), of the quadric, from the point $(0, 0, 1, -1)$; this is a point of the quadric, being one of the points in which it is met by the line given by the two equations $x = 0$, $y = 0$. It is clear that, while any pair (ξ, η) gives rise only to one point (x, y, z), this point, being the same as $(-x, -y, -z)$, gives rise, conversely, beside (ξ, η), to the pair

$$\xi' = \frac{-2x}{-z + t}, \quad \eta' = \frac{-2y}{-z + t},$$

which are $\xi' = -4\xi/(\xi^2+\eta^2)$, $\eta' = -4\eta/(\xi^2+\eta^2)$. Interpreting $(\xi, \eta, 2)$ as coordinates in a plane, so that the equation $\xi^2+\eta^2 = 4$ represents a circle, when (ξ, η) is within this circle, then (ξ', η') is without it, being such that $(-\xi, -\eta')$ and (ξ, η) are inverse points in regard to this circle (see pp. 67, 109, above). While, therefore, the complete plane (ξ, η) corresponds to all the points of the quadric $x^2 + y^2 + z^2 = t^2$, the complete plane (x, y, z) is represented by restricting ourselves, for example, to the interior of the circle $\xi^2+\eta^2 = 4$ (together with, say, the points of the circumference of this circle for which η is positive).

With these formulae, any line of the original plane, with the equation $ux + vy + wz = 0$, becomes, in terms of ξ and η, the locus represented by the equation

$$u\xi + v\eta + w\left[1 - \tfrac{1}{4}(\xi^2 + \eta^2)\right] = 0,$$

which, if $(\xi, \eta, 2)$ be coordinates, represents a circle cutting at right angles the fixed (imaginary) circle whose equation is $1+\tfrac{1}{4}(\xi^2+\eta^2) = 0$. Any such circle meets the real circle given by the equation $\xi^2+\eta^2-4 = 0$ at the ends of a diameter of this, and is, negatively, self-inverse in regard to this circle. Two such circles have two intersections of which only one is interior to the circle

$$\xi^2 + \eta^2 - 4 = 0.$$

Further, if (x', y', z') be any other point of the original plane, and we introduce corresponding symbols, ξ', η', and λ', μ', the interval, θ, between the points (x, y, z) and (x', y', z'), relatively to the conic $x^2 + y^2 + z^2 = 0$, which, as we have seen, is given by

$$\cos\theta = \frac{xx' + yy' + zz'}{tt'},$$

is equally such that

$$\sin^2\tfrac{1}{2}\theta = \frac{(\lambda'-\lambda)(\mu'-\mu)}{(1+\lambda\mu)(1+\lambda'\mu')},$$

as we at once verify. In terms of ξ, η and ξ', η', this leads to

$$\sin\tfrac{1}{2}\theta = D/PP',$$

where

$$D^2 = \tfrac{1}{4}\left[(\xi'-\xi)^2+(\eta'-\eta)^2\right], \quad P^2 = 1+\tfrac{1}{4}(\xi^2+\eta^2), \quad P'^2 = 1+\tfrac{1}{4}(\xi'^2+\eta'^2).$$

If we suppose $\lambda'-\lambda$, $\mu'-\mu$ to be both small, and replace θ by ds, we thence obtain, for the determination of the interval between the points (ξ, η) and $(\xi+d\xi, \eta+d\eta)$, the form

$$ds^2 = \frac{d\xi^2 + d\eta^2}{[1 + \tfrac{1}{4}(\xi^2 + \eta^2)]^2}.$$

This may, of course, be obtained directly from the value of ds^2 for the interval between the points (x, y, z) $(x + dx, y + dy, z + dz)$; this is at once found to be

$$ds^2 = \frac{(ydz - zdy)^2 + (zdx - xdz)^2 + (xdy - ydx)^2}{(x^2 + y^2 + z^2)^2}$$

$$= \frac{dx^2 + dy^2 + dz^2 - dt^2}{t^2}, \quad = \left[d\left(\frac{x}{t}\right)\right]^2 + \left[d\left(\frac{y}{t}\right)\right]^2 + \left[d\left(\frac{z}{t}\right)\right]^2.$$

The value obtained for ds, in terms of ξ, η, is precisely that found by Riemann (*loc. cit.* p. 264) for the element of length appropriate to a manifoldness which, in his sense, is of constant curvature, given by the absolute coordinates ξ, η. His result is for a metrical manifoldness of any number, n, of absolute coordinates. But it may be shewn, reversing the procedure here followed, that any such manifoldness is in correspondence with a manifoldness to which $(n + 1)$ homogeneous coordinates are appropriate; and that his measure of length is then obtained by considering the interval between two points with respect to an enveloping cone of this, as in the preceding work, where $x^2 + y^2 + z^2 = 0$ is the equation of the cone formed by the lines drawn from $(0, 0, 0, 1)$ to touch the quadric $x^2 + y^2 + z^2 = t^2$. The intermediate parameters λ, μ, which are convenient, but not necessary, are not available in the general case.

If in the formula for $\sin^2\frac{1}{2}\theta$, in terms of $\lambda, \mu, \lambda', \mu'$, we replace λ, μ, respectively, by $-\mu^{-1}, -\lambda^{-1}$, and make the corresponding replacements for λ', μ', the form remains the same. It also remains unchanged if we replace λ and $-\mu^{-1}$, respectively, by $(A\lambda + B)/(C\lambda + D)$ and $(-A\mu^{-1} + B)/(-C\mu^{-1} + D)$, wherein A, B, C, D are arbitrary constants. This last change corresponds to a linear transformation of the coordinates x, y, z which is such that the conic $x^2 + y^2 + z^2 = 0$ remains unchanged. (See p. 169.) The former change, in terms of ξ, η, is of the form $\xi' = -4\xi/(\xi^2 + \eta^2)$, $\eta' = -4\eta/(\xi^2 + \eta^2)$, which has been noticed above. The expression for $\sin^2\frac{1}{2}\theta$ is thus unaltered by a transformation obtained by combining these two transformations.

Further, it is easy to compute that, if we measure the angle between two circles, of equations

$$u\xi + v\eta + w\left[1 - \tfrac{1}{4}(\xi^2 + \eta^2)\right] = 0, \quad u'\xi + v'\eta + w'\left[1 - \tfrac{1}{4}(\xi^2 + \eta^2)\right] = 0,$$

by taking the tangents at one of their common points, (ξ_0, η_0), with respect to the two lines whose equations are $\xi - \xi_0 \pm i(\eta - \eta_0) = 0$, as explained above, then this angle between the circles is equal to the angular interval between the lines, of the plane (x, y, z), which correspond to these circles, this last interval being taken relatively to the conic $x^2 + y^2 + z^2 = 0$. That this must be so is obvious geometrically, as in the preceding section.

And, it may be verified, by the formulae $X = \xi [1 - \frac{1}{4}(\xi^2 + \eta^2)]^{-1}$, $Y = \eta [1 - \frac{1}{4}(\xi^2 + \eta^2)]^{-1}$, that the integral introduced above (p. 187), to measure the extent of a region, becomes, in terms of ξ, η,

$$\iint \frac{d\xi \, d\eta}{[1 + \frac{1}{4}(\xi^2 + \eta^2)]^2}.$$

This is unaltered by the substitutions

$$\xi' = - 4\xi/(\xi^2 + \eta^2), \quad \eta' = - 4\eta/(\xi^2 + \eta^2).$$

When applied to a region which does not lie wholly within the circle $\xi^2 + \eta^2 = 4$, it is to be taken only over that part of the region which does lie within this circle.

The case when the equation of the quadric surface is of the form $z^2 - x^2 - y^2 = t^2$ is similar to the preceding. It is convenient, however, in this case, to make a substitution of the form

$$x = \lambda + \mu, \quad y = 1 - \lambda\mu, \quad z = 1 + \lambda\mu, \quad t = -i(\lambda - \mu);$$

if we take $\lambda = \frac{1}{2}(\xi + i\eta)$, $\mu = \frac{1}{2}(\xi - i\eta)$, this gives $\xi = 2x/(y + z)$, $\eta = 2t/(y + z)$, and $x = \xi$, $y = 1 - \frac{1}{4}(\xi^2 + \eta^2)$, $z = 1 + \frac{1}{4}(\xi^2 + \eta^2)$, $t = \eta$.

By these formulae, any point of the original plane (x, y, z), with the associated positive value of t, is made to correspond to a point of the plane in which $(\xi, \eta, 2)$ are coordinates, only the points of this latter plane for which η is positive being considered. The separating line $\eta = 0$ plays, however, a different part from that played by the circle $\xi^2 + \eta^2 = 4$ in the former case: the line $\eta = 0$ consists of points each corresponding only to one point of the quadric, and is a natural boundary; the circle $\xi^2 + \eta^2 = 4$ was generated as a locus of pairs of points, and was a conventional boundary, convenient but not unique. Further, the formulae give $z^2 - x^2 - y^2 = \eta^2$; thus points of the original plane (x, y, z) which are outside the conic $z^2 - x^2 - y^2 = 0$, as they would correspond to a negative value of η^2, are not represented by real points of the plane $(\xi, \eta, 2)$; the curves of the plane $(\xi, \eta, 2)$ corresponding to two lines of the plane (x, y, z), which do not intersect within the conic $z^2 - x^2 - y^2 = 0$, will be curves with no real intersection. In fact a line whose equation is $ux + vy + wz = 0$ is represented in the plane $(\xi, \eta, 2)$ by the locus whose equation is

$$u\xi + v[1 - \frac{1}{4}(\xi^2 + \eta^2)] + w[1 + \frac{1}{4}(\xi^2 + \eta^2)] = 0;$$

this is a circle whose centre is on the line $\eta = 0$, of which, as we have said, we consider only the points for which η is positive. All such circles arise, if all lines of the original plane be considered, and conversely. When two such circles intersect in two points, only one will be such that η is positive; this will correspond to an intersection of two lines, in the original plane, at a point interior to the conic $z^2 - x^2 - y^2 = 0$; when two such circles meet (and therefore touch) at a point for which $\eta = 0$, they correspond to lines of the original plane which meet on the absolute conic, and may, therefore, be said to be parallel. It is easy to see that, through an arbitrary point, of the plane $(\xi, \eta, 2)$, for which η is positive, there can be drawn two circles, with centres on the line $\eta = 0$, each of which shall touch (on $\eta = 0$) a given circle whose centre is on $\eta = 0$. Particular (degenerate) circles of the system are the lines, of equations of the form $\xi = $ constant, which are at right angles to the line $\eta = 0$; these correspond to the lines of the original plane which meet at the point $(0, 1, - 1)$ of the conic $z^2 - x^2 - y^2 = 0$. It is easy to compute that the angle between any two circles of the system, estimated as before by considering the tangents at the intersection, (ξ_0, η_0), of the circles, with respect to the two lines

$$\xi - \xi_0 \pm i\,(\eta - \eta_0) = 0,$$

is equal to the angular interval between the lines of the original plane which they represent, estimated with respect to the conic $z^2 - x^2 - y^2 = 0$. For the case of this conic, instead of the original interval, θ, between two points, we introduced an interval σ such that $\theta = i\sigma$. It can be shewn that, for the plane $(\xi, \eta, 2)$, this is such that

$$\sinh^2 \tfrac{1}{2}\sigma = \frac{(\lambda' - \lambda)(\mu' - \mu)}{-(\lambda - \mu)(\lambda' - \mu')}, \quad = \frac{(\xi' - \xi)^2 + (\eta' - \eta)^2}{4\eta\eta'},$$

becoming infinite when one of the points is on $\eta = 0$. When $\xi' = \xi + d\xi$, $\eta' = \eta + d\eta$, this gives

$$d\sigma = \frac{(d\xi^2 + d\eta^2)^{\frac{1}{2}}}{\eta};$$

and, by the ordinary methods of the Calculus of Variations, it will be found that the curve for which the integral

$$\int (d\xi^2 + d\eta^2)^{\frac{1}{2}}/\eta$$

taken between two fixed points, has a stationary value, is the circle through these points whose centre is on the line $\eta = 0$. We can interpret the interval σ, between two points P, P', by drawing, through these, the circle whose centre is on $\eta = 0$, and considering the points P, P' with respect to the two points, say M and N, in which this circle meets the line $\eta = 0$. For the pencil which these four points subtend at any point of the circle we then have, supposing that N, P, P', M are in order upon the portion of the circle considered, for which η is positive, the equation (p. 166)

$$(P, P'; M, N) = e^{-\sigma}.$$

The curve, called by Lobatschewsky a circle, of centre (x', y', z'), whose equation is (above, p. 182)

$$(zz' - xx' - yy')^2 - \lambda^2 (z^2 - x^2 - y^2) = 0,$$

where λ is supposed real and positive, is represented, in the part of

the $(\xi, \eta, 2)$ plane for which η is positive, by the circle whose
equation is

$$(\xi - \xi')^2 + (\eta - \lambda)^2 = \lambda^2 - \eta'^2 \, ;$$

this circle has its centre at $\xi = \xi'$, $\eta = \lambda$. It touches the line $\eta = 0$
and becomes a horocycle if $\eta' = 0$, that is, if
(x', y', z') lie on $z^2 - x^2 - y^2 = 0$.

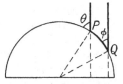

By means of any two points P, Q, a region
is defined, bounded by the lines through P, Q
with equations of the form $\xi = \text{constant}$, and
by the circle through P, Q whose centre is on
$\eta = 0$, but lying entirely outside this circle.
If θ, ϕ be the angles between this circle and the two lines spoken
of, respectively at P and Q, this being measured in the Euclidean
way, by means of lines $\xi - \xi_0 \pm i(\eta - \eta_0) = 0$, it is easy to shew, if
$\theta - \phi$ be positive, that the integral $\iint \eta^{-2} d\xi \, d\eta$, extended over this
region, has the value $\theta - \phi$.

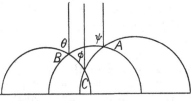

Suppose now that we have
three circles, with centres on
$\eta = 0$, whose intersections in
pairs are A, B, C. Any two of
these points, say A and B, determine such a region as we
have just considered, which we
may denote momentarily by (A, B). The points A, B, C determine
a triangle, whereof the joins are arcs of these circles; and, to limit
ourselves to one case, this triangle may be supposed to be obtained
by the addition of the two regions (B, C) and (C, A) and the subsequent subtraction of the region (B, A). With appropriate angles
θ, ϕ, ψ (explained by the annexed diagram), the integral $\iint \eta^{-2} d\xi \, d\eta$,
taken over the triangle, is then equal to

$$[\theta - \phi] + [(\pi - C + \phi) - (A + \phi)] - [(B + \theta) - \psi]$$

which is $\pi - A - B - C$, where A, B, C are the measures of the
angles of the triangle. We have, above (p. 180), given a proof that
this expression is necessarily positive.

The forms for x, y, z in terms of ξ, η,
suggest, putting $X = x/z$, $Y = y/z$, that we
consider the transformation

$$X = \xi[1 + \tfrac{1}{4}(\xi^2 + \eta^2)]^{-1},$$

$$Y = [1 - \tfrac{1}{4}(\xi^2 + \eta^2)] [1 + \tfrac{1}{4}(\xi^2 + \eta^2)]^{-1} \, ;$$

it is found that, then, the above integral,
$\iint \eta^{-2} d\xi \, d\eta$, becomes

$$\iint (1 - X^2 - Y^2)^{-\frac{3}{2}} dX \, dY.$$

Thus we infer that if, in the original plane (x, y, z), the point

(0, 1, −1) be T, and P, Q be two points of any line, interior to the absolute conic, so that TPQ is a triangle for which the angular interval between TP and TQ is zero, then this integral, extended over the triangle TPQ, is equal to the difference of the angular intervals between PQ and, respectively, PT and QT.

Comparison of the two cases considered. The distinction between the two cases considered above, of an imaginary conic of reference, and a real conic, has arisen by respecting the distinction between real and imaginary points. If, in the former case, we put $t_1 = -it$, and, in the latter case, put $z_1 = -iz$, as well as λ', μ' in place of λ, μ, the formulae are

$$x = \lambda + \mu, \quad y = -i(\lambda - \mu), \quad z = 1 - \lambda\mu, \qquad t_1 = -i(1 + \lambda\mu),$$
$$x = \lambda' + \mu', \quad y = 1 - \lambda'\mu', \quad z_1 = -i(1 + \lambda'\mu'), \quad t = -i(\lambda' - \mu'),$$

and the equations of the quadric surface, in these two cases, are $x^2 + y^2 + z^2 + t_1^2 = 0$ and $x^2 + y^2 + z_1^2 + t^2 = 0$, respectively. If we now replace z_1, t_1, respectively, by z and t, the comparison of the formulae leads to

$$\lambda' = -i\frac{1+\lambda}{1-\lambda}, \quad \mu' = i\frac{1-\mu}{1+\mu}.$$

Corresponding to the twenty-four orders in which the four coordinates x, y, z, t may be taken, there follow twenty-four such ways in which we may replace x, y, z, t by functions of two parameters λ, μ, so as to satisfy, identically, the equation of the quadric surface. These twenty-four sets of formulae can all be obtained from any one of them by replacing λ and μ by proper linear (fractional) functions, respectively of λ and μ, or of μ and λ. It is not difficult to see that these twenty-four pairs of replacements are combinations of the three pairs

$$(\alpha), \ \lambda' = \lambda^{-1}, \ \mu' = -\mu \ ; \quad (\beta), \ \lambda' = i\mu, \ \mu' = -i\lambda \ ;$$
$$(\gamma), \ \lambda' = (1 - \mu)(1 + \mu)^{-1}, \quad \mu' = (1 - \lambda)(1 + \lambda)^{-1} \ ;$$

for instance the replacement above is $\alpha\beta\gamma$, obtained by first carrying out γ, then β, and then α. Neglecting the distinction between real and imaginary, we thus have, by putting $\lambda = \frac{1}{2}(\xi + i\eta)$, $\mu = \frac{1}{2}(\xi - i\eta)$, twenty-four such representations of the original plane (x, y, z), upon a plane (ξ, η).

Examples of the preceding considerations. *Ex.* 1. *Delambre's formulae in spherical trigonometry.* Let the absolute conic, which we suppose to be an imaginary conic, when referred to three real points X, Y, Z, have the equation

$$ax^2 + 2fyz + \ldots = 0,$$

containing six terms, of which the coefficients, $a, f, \ldots$, are real. No one of a, b, c is zero, and no one of $bc - f^2$, $ca - g^2$, $ab - h^2$, for the (imaginary) conic does not contain any of the points X, Y, Z, nor touch any of the lines YZ, ZX, XY. As usual, let $A = bc - f^2$, and $F = gh - af$, etc., and $\Delta = abc + 2fgh - af^2 - bg^2 - ch^2$. The left side of the equation being

$$a \left(x + \frac{h}{a} y + \frac{g}{a} z \right)^2 + \frac{C}{a} \left(y - \frac{F}{C} z \right)^2 + \frac{\Delta}{C} z^2,$$

it follows that a, C, Δ, and, therefore, similarly, all of b, c, A, B, may be taken to be positive (and not zero). We can then define intervals, YZ, ZX, XY, which we denote, respectively, by α, β, γ, and also angular intervals between the pairs of these lines, which we denote by α', β', γ', the first being that between XY and XZ, and so on, by the equations (following from pp. 175, 176)

$$\cos \alpha = \frac{f}{(bc)^{\frac{1}{2}}}, \qquad \cos \beta = \frac{g}{(ca)^{\frac{1}{2}}}, \qquad \cos \gamma = \frac{h}{(ab)^{\frac{1}{2}}},$$

$$\sin \alpha = \left(\frac{A}{bc} \right)^{\frac{1}{2}}, \qquad \sin \beta = \left(\frac{B}{ca} \right)^{\frac{1}{2}}, \qquad \sin \gamma = \left(\frac{C}{ab} \right)^{\frac{1}{2}},$$

$$\cos \alpha' = \frac{-F}{(BC)^{\frac{1}{2}}}, \qquad \cos \beta' = \frac{-G}{(CA)^{\frac{1}{2}}}, \qquad \cos \gamma' = \frac{-H}{(AB)^{\frac{1}{2}}},$$

$$\sin \alpha' = \left(\frac{a\Delta}{BC} \right)^{\frac{1}{2}}, \qquad \sin \beta' = \left(\frac{b\Delta}{CA} \right)^{\frac{1}{2}}, \qquad \sin \gamma' = \left(\frac{c\Delta}{AB} \right)^{\frac{1}{2}};$$

and it is then easy to see, comparing the relations previously obtained (above, p. 177), that these belong to a triangle of which X, Y, Z are the vertices. It is understood that, for example, $(bc)^{\frac{1}{2}}$ means $b^{\frac{1}{2}} \cdot c^{\frac{1}{2}}$, both the factors being real, and so in each case. The signs of the square roots are to be determined consistently with the formulae referred to.

Now, if ϵ be $+1$ or -1, and σ be $+1$ or -1, we have, identically,

$$\frac{A\left[(BC)^{\frac{1}{2}} + \epsilon F \right]}{\Delta \left[(bc)^{\frac{1}{2}} - \sigma f \right]} = \frac{a \left[(bc)^{\frac{1}{2}} + \sigma f \right]}{(BC)^{\frac{1}{2}} - \epsilon F};$$

hence, as $AF = GH - f\Delta$, or directly, we see that the function

$$\left\{ 1 + \epsilon \frac{G}{(CA)^{\frac{1}{2}}} \frac{H}{(AB)^{\frac{1}{2}}} - \sigma\epsilon \left(1 - \frac{G^2}{CA} \right) \left(1 - \frac{H^2}{AB} \right)^{\frac{1}{2}} \right\} \div \left\{ 1 + \epsilon \frac{F}{(BC)^{\frac{1}{2}}} \right\}$$

is equal to the function obtained from this by replacing A, B, C, F, G, H, respectively, by a, b, c, f, g, h, and interchanging ϵ and σ. Wherefore

$$\{ 1 + \epsilon \cos \beta' \cos \gamma' - \sigma\epsilon \sin \beta' \sin \gamma' \} \div \{ 1 - \epsilon \cos \alpha' \}$$

is equal to

$$\{ 1 + \sigma \cos \beta \cos \gamma - \sigma\epsilon \sin \beta \sin \gamma \} \div \{ 1 + \sigma \cos \alpha \}.$$

For instance, when $\epsilon = \sigma = 1$, we have

$$\left\{\frac{\cos \frac{1}{2}(\beta' + \gamma')}{\sin \frac{1}{2}\alpha'}\right\}^2 = \left\{\frac{\cos \frac{1}{2}(\beta + \gamma)}{\cos \frac{1}{2}\alpha}\right\}^2,$$

and similarly in the other cases.

Ex. 2. Napier's analogies. Let A, B, C be a triangle, in which

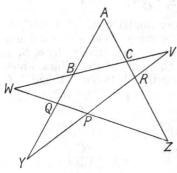

AB, AC are conjugate lines, in regard to an imaginary conic of reference, so that the angular interval BAC is $\frac{1}{2}\pi$. Let the polar of B meet AB, AC, BC, respectively, in Y, R and V, and the polar of C meet AC, AB, BC, respectively, in Z, Q and W, while these polars meet in P, which is thus the pole of BC.

It can then be shewn that, as at A, the pairs of lines meeting in Y, Z, V, W are, in each of these four cases, conjugate to one another; and that, in the pentagon $PRCBQ$, each vertex is the pole of the opposite side.

Each of the four triangles CRV, BQW, PQY, PRZ has its elements, three segments and three angles, expressible in a simple way in terms of those of the original triangle A, B, C. Obtain these expressions. Hence shew, in particular, that the relation $\sin RV = \sin CR \sin (VCR)$, which holds for the triangle CRV (see above, p. 176), involves, for the triangle ABC, the relation

$$\cos B = \cos b \sin C;$$

similarly, shew that the relation $\sin BW = \sin BQ \sin (BQW)$ in-

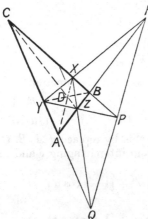

volves, for ABC, the relation

$$\cos a = \cos b \cos c;$$

and so on. (John Napier, *Mirifici logarithmorum canonis descriptio*, Edinburgh, 1614, *Lib.* ii, *Cap.* iv. Compare Gauss, *Ges. Werke*, iii, p. 481.)

Ex. 3. A particular triangle. Let X, Y, Z be a self-polar triad, of real points, in regard to the imaginary conic $x^2 + y^2 + z^2 = 0$. Let any line, PQR, meet YZ, ZX, XY, respectively, in P, Q, R, the lines YQ, ZR meeting in A, the lines ZR, XP in B, and the lines XP, YQ in C. The lines XA, YB, ZC will then meet in a point, say D. If the equation of PQR be

$a^{-1}x + b^{-1}y + c^{-1}z = 0$, the coordinates of A, B, C, D will respectively be $(-a, b, c)$, $(a, -b, c)$, $(a, b, -c)$, (a, b, c). Let d be the positive quantity $(a^2 + b^2 + c^2)^{\frac{1}{2}}$. In regard to $x^2 + y^2 + z^2 = 0$, the pole of the line XP lies in that segment YZ which does not contain P; the line through B, at an angular interval with the line BXC equal to $\frac{1}{2}\pi$, thus contains a point of this segment; so that the angular interval between BXC and BAR is greater than $\frac{1}{2}\pi$. Similarly the angular interval between AYC and ABR is greater than $\frac{1}{2}\pi$, and the angular interval between CXB and CYA is greater than $\frac{1}{2}\pi$.

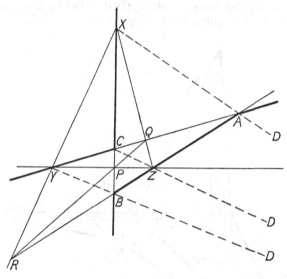

This is true for all positions of P, Q, R in the segments YZ, ZX, XY. We may consider two cases, that in which a, b, c are all positive, and that in which a, b are positive, but c is negative. Let the angular intervals referred to, at A, B, C, be denoted, respectively, by α', β', γ'. We then have (see above, p. 174), the equations of XP, YQ, ZR being, respectively, $b^{-1}y + c^{-1}z = 0$, $c^{-1}z + a^{-1}x = 0$, $a^{-1}x + b^{-1}y = 0$, for the case in which a, b, c are all positive,

$$\cos \alpha' = -bc\,(a^2 + b^2)^{-\frac{1}{2}}(a^2 + c^2)^{-\frac{1}{2}},$$

and similar expressions, $\sin \alpha' = ad\,(a^2 + b^2)^{-\frac{1}{2}}(a^2 + c^2)^{-\frac{1}{2}}$, etc.

From these we find $\cos \alpha' = \cos(\beta' + \gamma')$, and similar equations, and infer that $\alpha' + \beta' + \gamma' = 2\pi$. (Cf. p. 197.)

The same follows when c is negative; the values of $\cos \alpha'$, $\cos \beta'$, $\sin \gamma'$ are then changed in form, by the substitution of $-c$ for c.

If α, β, γ be the respective intervals BC, CA, AB for the segments which are appropriate to α', β', γ', in order to form a triangle, according to the definition above given (p. 172), it follows from the formulae such as $\cos\alpha \sin\beta' \sin\gamma' = \cos\alpha' + \cos\beta' \cos\gamma'$, that $\cos\alpha = (a^2 - b^2 - c^2).d^{-2}$, $\cos\beta = (b^2 - c^2 - a^2).d^{-2}$, $\cos\gamma = (c^2 - a^2 - b^2).d^{-2}$, while $\sin\alpha = 2a(b^2 + c^2)^{\frac{1}{2}} d^{-2}$, and so on, the form for $\sin\gamma$ being $-2c(a^2 + b^2)^{\frac{1}{2}} d^{-2}$ when c is negative. Of the expressions, $a^2 - b^2 - c^2$, $b^2 - c^2 - a^2$, $c^2 - a^2 - b^2$, two at least must be negative, as we easily prove, remarking that the sum of every two is negative; evidently, $1 + \cos\alpha + \cos\beta + \cos\gamma$ is zero. It is interesting to remark also that if the angular intervals BCA, CAB, ABC be measured with reference to the two points $z = 0$, $x \pm iy = 0$, the sum of these intervals

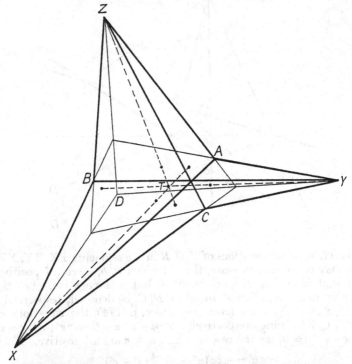

is also 2π, as follows from the fact that the interval YCX is then $\frac{1}{2}\pi$ (cf. p. 167 preceding). The figure can be obtained by taking, upon the quadric $x^2 + y^2 + z^2 = t^2$, the points $(-a, b, c, d)$, $(a, -b, c, d)$, $(a, b, -c, d)$, and $(-a, -b, -c, d)$, and projecting from the point $(0, 0, 0, 1)$ upon the plane XYZ, given by $t = 0$. The planes TAB, TAC intersect in the line ATA', where A' is $(a, -b, -c, d)$, and are at an angular interval α'. Upon the quadric the four triangles

ABC, *BCD*, *CAD*, *ABD* have, every two, equal angles and seg-ments (p. 197).

The relations for the angles and segments of the triangle *ABC* may be obtained also by remarking (see above, Chap. III, Ex. 14) that the tangents to the absolute conic from *A*, *B*, *C* meet, respec-tively, the lines *BC*, *CA*, *AB*, in points which lie in threes upon two lines. If we put $\xi = b^{-1}y + c^{-1}z$, $\eta = c^{-1}z + a^{-1}x$, $\zeta = a^{-1}x + b^{-1}y$, the equation of the absolute conic referred to *A*, *B*, *C* is

$$a^2(-\xi + \eta + \zeta)^2 + \ldots = 0, \text{ or } \xi^2 + 2\cos\alpha \cdot \eta\zeta + \ldots = 0,$$

whose tangential form is $\sin^2\alpha \cdot u^2 - 2\sin\beta \sin\gamma \cos\alpha' \cdot vw + \ldots = 0$. The condition in question, writing *l*, *m*, *n* respectively for $u \sin\alpha$, $v \sin\beta$, $w \sin\gamma$, is, that there should be a line (*l*, *m*, *n*) for which each of the equations $m^2 + n^2 - 2mn \cos\alpha' = 0$, $n^2 + l^2 - 2nl \cos\beta' = 0$, $l^2 + m^2 - 2lm \cos\gamma' = 0$ is satisfied ; which is so if $\alpha' + \beta' + \gamma' = 2\pi$.

Ex. 4. *General form of the theorem for the sum of the focal dis-tances of a point of a conic.* If the common tangents of any conic, Ω, and the absolute conic, be taken, and *S*, *H* be a point pair through which all these tangents pass, we may prove that the intervals *PS*, *PH*, in regard to the absolute conic, for any point *P* of the conic Ω, have a constant sum (or difference). When the absolute conic is an imaginary point pair, this gives the familiar theorem of metrical geometry, for the sum (or difference) of the focal distances of a point of a conic.

This result may be deduced from the following : Let *A*, *B*, *C*, *D* be four variable points of the absolute conic, such that *AB* passes through a fixed point, *S*, while *BC* passes through a fixed point, *M*, and *CD* through a fixed point, *H*, the points *S*, *H*, *M* being in line. Then *DA* passes through a further fixed point, *N*, of this line. And the locus of the intersection, *P*, of the lines *AB*, *CD*, is a conic, which touches the tangents to the absolute conic from the points *S* and *H*, and passes through the points of contact of this absolute conic with the tangents drawn to it from *M* and *N*.

Although an anticipation, it seems proper to mention, in con-nexion with this, two theorems given by Chasles, *Géom. Supér.* 1880, pp. 517, 528. If the points of the conic Ω be joined by lines to a point *O* not lying in the plane of the conic, the points of these lines are said to lie on a quadric cone ; the points of the cone lying in any plane are points of a conic. If any plane be drawn through a common chord of Ω and the absolute conic, the conic in which this plane meets the cone is called a circular section of the cone. A plane through the complementary common chord of the two conics meets the cone in a circular section complementary to any one of the former system. Now let the angle between any two planes be measured by the interval between the lines in which these planes meet the plane of the absolute conic, in respect to this conic. Then we have the result that the sum (or difference) of the angles between any tangent plane of the cone and a pair of complementary circular sections, is constant. Again, if O be joined to one of the point pairs through which pass the common tangents of the conic Ω and the absolute conic, the two joining lines are called a pair of focal lines of the cone. If the angle between two lines, not lying in the plane of the absolute conic, be measured by the interval between the points in which these meet the plane of the absolute conic, taken with respect to this, we have the result that the angles which any edge of the cone makes with a pair of focal lines have a constant sum (or difference).

Ex. 5. If on a conic there be taken, arbitrarily, the points

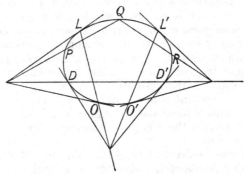

D, D', O, O', P, and then there be determined in succession, L, L', Q, R, by means of $(LO, DD') = -1$, $(L'O', DD') = -1$, $(QP, OL) = -1$, $(RQ, O'L') = -1$, so that L is the harmonic conjugate of O, on the conic, in regard to D and D', and so on, prove that (RP, DD') is equal to $(O'O, DD')^2$.

Ex. 6. If the tangents at the points C, D of a conic meet in F,

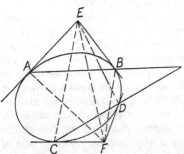

and the tangents at the points A, B meet in E, prove that, measured with respect to this conic, the angular intervals AFE, EFB, CEF, FED are equal. Denoting this interval by α, and the angular interval between the lines AB and CD by β, prove that $\sin \alpha \sin \beta = \pm 1$.

Ex. 7. If PQ, PR be two tangents to a conic which touches the absolute conic in two points, the intervals PQ, PR are equal; and are the same as for any other point, P', lying on a conic through P which touches the absolute conic at the same two points as the former.

Ex. 8. Let V and W be two conics each of which touches the absolute conic in two points, not necessarily the same, and U be a conic touching each of V and W in two points; from any point, P, of U let a tangent PQ be drawn to V, and a tangent PR be drawn to W; prove that the intervals PQ, PR have a constant sum (or difference) as P varies on U. Shew that this includes the result given in Ex. 4.

Ex. 9. From a point, P, of a conic $f = 0$, a tangent PT is drawn to a conic $\psi = 0$, and PT meets $f = 0$ again in P'. Let $f_1 = \partial f/\partial x$, etc., and c_1, c_2, c_3 be arbitrary. Prove that the differential, taken along the conic $f = 0$, at P, which is expressed by

$$\begin{vmatrix} dx, & dy, & dz \\ x, & y, & z \\ c_1, & c_2, & c_3 \end{vmatrix} \div (c_1 f_1 + c_2 f_2 + c_3 f_3)(\psi)^{\frac{1}{2}},$$

has the same value as the corresponding differential at P'.

The conics being in general relation, the form of the integral shews that there is no loss of generality in supposing $\psi = \xi^2 + \eta^2 - 1$, $f = a\xi^2 + b\eta^2 - 1$. Let $F(\xi)$ denote what f becomes by putting $\eta = \operatorname{cosec} \alpha . (1 - \xi \cos \alpha)$, and $F'(\xi) = \partial F/\partial \xi$; by this substitution we have $\partial \eta/\partial \alpha = \operatorname{cosec}^2 \alpha . (\xi - \cos \alpha) = \pm \operatorname{cosec} \alpha . \psi^{\frac{1}{2}}$. From $F(\xi) = 0$, the integral $- d\xi/\eta\psi^{\frac{1}{2}}$ becomes $2b . d\alpha . [F'(\xi) . \psi^{\frac{1}{2}}]^{-1} \partial \eta/\partial \alpha$, or

$$\pm 2b . d\alpha . \operatorname{cosec} \alpha/F'(\xi);$$

and $F'(\xi)$ has opposite signs at P and P'.

Similarly for other forms of f. (Cf. Ex. 25, p. 147.)

NOTE I

ON CERTAIN ELEMENTARY CONFIGURATIONS, AND ON THE COMPLETE FIGURE FOR PAPPUS' THEOREM

The present note deals in an incomplete way with certain configurations which present themselves naturally, especially in the consideration of the complete figure for Pascal's theorem which is considered in the succeeding note.

1. The simpler Desargues' figure. If A,B,C and A',B',C' be two triads of points in perspective, from a point O, and the lines $BC, B'C'$ meet in L, while $CA, C'A'$ meet in M, and $AB, A'B'$ meet in N, then L, M, N lie in line.

Regarded as a plane figure, there are here, in all, ten points,

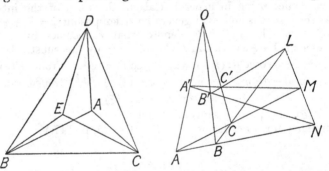

each lying on three lines, and ten lines, each containing three points; and the dual of the figure is of the same character. To express the incidences it may be denoted by $10_3 10_3$ (or simply 10_3), or by $10(. 3) . 10(3 .)$.

Regarded as a figure in three dimensions, the figure is formed by the intersections of five planes, the ten points being the intersections of threes of these planes, and the ten lines the intersections of twos of these planes. Each point now lies in three lines and in three planes, each line contains three points and lies in two planes, and each plane contains six points and four lines. These incidences we may summarise by denoting the figure by

$$10 (. 3, 3) . 10 (3 . 2) . 5 (6, 4 .).$$

The figure, however the points be taken, necessarily lies in a threefold space. Its dual, in this space, is formed by five points, the lines joining the pairs of these, and the planes containing the threes of these. But this dual figure may evidently be regarded as obtained by projection from a figure formed by taking five points in a space of four dimensions. For this figure, we shall naturally take into account, beside the lines and planes, also the threefold spaces defined by the fours of the five given points. The representation of the incidences will then be by a symbol

$$5 (. 4, 6, 4) . 10 (2 . 3, 3) . 10 (3, 3 . 2) . 5 (4, 6, 4 .).$$

And the dual of this figure in four dimensions will be a figure of the same character.

2. **The figure of three desmic tetrads of points in space of three dimensions.** Two triads of points, A, B, C and A', B', C', which are in perspective, enable us to define two sets each of four planes, namely ABC, $AB'C'$, $BC'A'$, $CA'B'$ and $A'B'C'$, $A'BC$, $B'CA$, $C'AB$; these planes have the property that the sixteen lines in which any one of the first four planes meets the second four planes, lie by fours in four other planes.

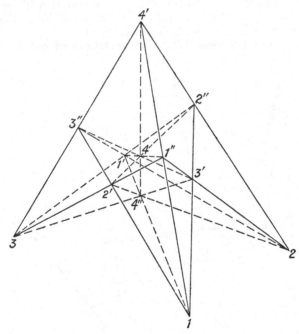

We consider the dual of this figure in three dimensions. We take six planes meeting in pairs in three lines which lie in a plane; namely the planes $234'$, $234''$; $314'$, $314''$; $124'$, $124''$; then the planes 314', 124' meet in a line, 14'; the intersection of this with the plane 234'' is a point, 1''; similarly the planes 314'', 124'' meet in a line, 14''; the intersection of this with the plane 234' is a point, 1'. Let the points 2'', 2' and 3'', 3' be similarly defined. It can be shewn that the lines 1'1'', 2'2'', 3'3'', 4'4'' meet in a point, 4. Thus the twelve points lie in threes on sixteen lines in the following way:

$$12'3'', \quad 12''3', \quad 11'4'', \quad 11''4',$$
$$23'1'', \quad 23''1', \quad 22'4'', \quad 22''4',$$
$$31'2'', \quad 31''2', \quad 33'4'', \quad 33''4',$$
$$41'1'', \quad 42'2'', \quad 43'3'', \quad 44'4'',$$

and any two of the four tetrads of points 1, 2, 3, 4; 1', 2', 3', 4'; 1'', 2'', 3'', 4'' are in perspective in four ways, the centres of perspective being the points of the other tetrad.

Beside the twelve points, and the sixteen lines, the figure contains twelve planes, consisting of the six from which we started, together with the three 14 2'3'2''3'', 24 3'1'3''1'', 34 1'2'1''2'', and the three 14 1'4'1''4'', 24 2'4'2''4'', 34 3'4'3''4''. The description of the complete figure is given by 12 (. 4, 6) 16 (3 . 3) 12 (6, 4.). The dual of this figure is a figure of the same character.

3. **The figure formed from three triads of points which**

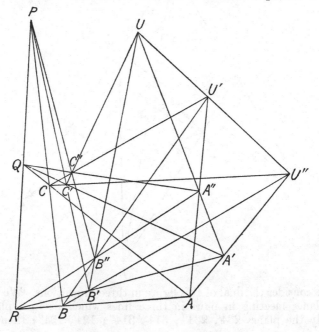

are in perspective in pairs, with centres of perspective in line. This figure, which is that used in the proof of Desargues' theorem for two triads in one plane (Vol. I, Frontispiece, and p. 7), may be described as consisting of two tetrads of points, Q, C, C', C'' and R, B, B', B'', in perspective from P, together with the six points of intersection of the corresponding joins in these tetrads, such as QC and RB, or $C'C''$ and $B'B''$, namely, the points A, A', A'', U, U', U'', which lie in threes in four lines lying in a plane. Any point of the figure may be taken, in place of P, as such a centre of perspective.

The figure may be in four dimensions, but not in higher dimensions. So considered it contains fifteen points, twenty lines and fifteen planes, namely the three planes, $ABC, A'B'C', A''B''C''$, passing through the line PQR, and three similar planes through the line $UU'U''$, containing, respectively, the triads $AA'A'', BB'B'', CC'C''$, together with nine other planes, of which for instance there are three through P, namely, $PB'C'B''C''$, which contains the point U; $PB''C''BC$, which contains the point U'; and $PBCB'C'$, which contains the point U''. The figure also contains six spaces of three dimensions. One of these, for instance, contains the ten points $P, Q, R, A, B, C, A', B', C', U''$; another contains the ten points $U, U', U'', B, B', B'', C, C', C'', P$; the former of these contains two planes through the line PQR, the two planes which join the line $AA'U''$ to Q and R, and the plane $PBB'CC'$, five in all, and is, in fact, such a figure as considered above, in § 1. The whole description of the incidences of the figure is then given by the scheme

$$15\,(\,.\,4, 6, 4)\ 20\,(3\,.\,3, 3)\ 15\,(6, 4\,.\,2)\ 6\,(10, 10, 5\,.\,).$$

If this figure be projected into a plane, it becomes the dual of a generalisation of the complete figure for Pappus' theorem. We shall however obtain this by taking a section, by a plane, of the dual of this figure in four dimensions, which will appear to have great simplicity. But first we consider the complete figure of Pappus' theorem.

4. The complete figure for Pappus' theorem. The points A, B, C being in line, and the points A', B', C' in another line intersecting the former, the pairs of lines $BC', B'C$; $CA', C'A$ and $AB', A'B$ intersect in three points lying on a line, say k_1. This is Pappus' theorem referred to. We may however, retaining the order of A, B, C, take A', B', C' in six different orders, and so obtain in all six lines such as k_1, each containing three points of intersection. It is found however that the three lines, k, corresponding to the orders A', B', C'; B', C', A' and C', A', B' meet in a point, say G; and the three lines, k, corresponding to the orders A', C', B';

B', A', C'; C', B', A' also meet in a point, say G'. This was re-marked by Steiner (*Werke*, I, pp. 451, 452). If now we omit the lines ABC, $A'B'C'$, and the points A, B, C, A', B', C', and add to

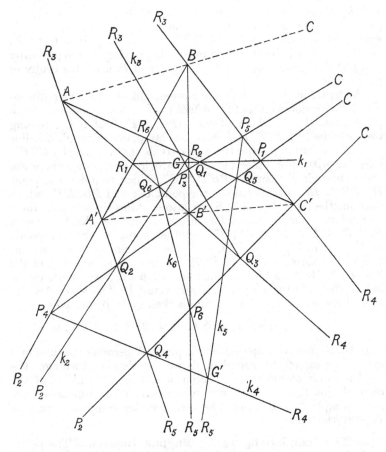

the points the two points G and G', we have a figure containing in all twenty points, each being an intersection of three lines, and containing fifteen lines (the nine lines originally arising as the joins of A, B, C with A', B', C', and the six lines k) upon each of which lie four of the points. The description of the incidences of the figure will then be $20\,(\,.\,3)\,15\,(4\,.\,)$.

5. **The figure formed from six points in four dimensions.** Consider in four dimensions the dual of the figure described in § 3. We may construct it by the dual procedure to that before employed;

but it will be simpler to take the scheme of incidences which we can infer from that for the other figure. This will be

$$6\,(\,.\,5,10,10)\;15\,(2\,.\,4,6)\;20\,(3,3\,.\,3)\;15\,(4,6,4\,.\,).$$

From this it can be seen at once that the figure consists of six points, the lines joining them in pairs, the planes containing threes of them, and the threefold spaces defined by fours of them.

Consider now this figure, and the section of it by an arbitrary plane, ϖ. The figure being in four dimensions, a line of it will not in general have any intersection with ϖ; a plane of it will meet ϖ in a point; and a threefold space of it will meet ϖ in a line. Thus, the lines in which ϖ is met by two threefold spaces of the figure will meet in the point in which ϖ is met by the plane in which the two threefold spaces intersect; and the points in which ϖ is met by three planes which belong to the same threefold space of the figure, lie in a line, in which this threefold space meets ϖ.

Though the figure is symmetrical, we shall grasp the relations of the plane figure better, perhaps, if we choose out, from the figure in four dimensions, two planes which have no point in common, and the three spaces which contain each of these planes. Denoting the six fundamental points by A, B, C and A', B', C', we take the planes ABC and $A'B'C'$, and the six threefold spaces

$$A',A,B,C;\quad B',A,B,C;\quad C',A,B,C,$$
$$A,A',B',C';\quad B,A',B',C';\quad C,A',B',C'.$$

Save for the planes ABC and $A'B'C'$, no two of these six spaces have a plane in common.

With each of these six spaces we can now associate six other spaces—as we explain immediately. But of the thirty-six spaces so derived, each occurs four times in the enumeration. The nine different ones, together with the original six, are the whole fifteen spaces of the figure.

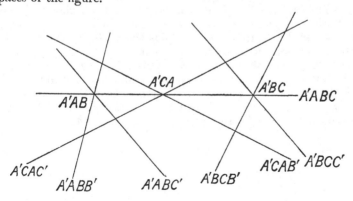

Take, for instance, the space A', A, B, C. Beside the plane ABC, this contains the three planes $A'BC, A'CA, A'AB$. Each of these lies in two other spaces beside $A'ABC$, namely, respectively,

	$A'BC$,	$A'CA$,	$A'AB$
lie in	$A'BCB'$,	$A'CAC'$,	$A'ABB'$
and in	$A'BCC'$,	$A'CAB'$,	$A'ABC'$.

We thus have, by section with the plane ϖ, a figure such as that occurring in Pappus' theorem, but in a generalised form. The two original triads of points of Pappus' theorem, each in line, are here replaced by triads of points which are not generally in line. The reader will easily find the specialisation of the six points sufficient to secure that they shall be.

Ex. 1. If two sets, each of n points, $A_1 \ldots A_n$ and $B_1 \ldots B_n$, in $(n-1)$ fold space, be in perspective, from a point O, there is an unique quadric $(n-2)$ fold, in regard to which any point, say A_1, of one set, is the pole of the $(n-2)$ fold containing the corresponding points, $B_2 \ldots B_n$, of the other. If C_{rs} be the intersection of the lines $A_r A_s$, $B_r B_s$, the points C_{rs} lie in a $(n-2)$ fold, which is the polar of O in regard to the quadric. Further, the figure is symmetrical, everyone of the $\frac{1}{2}(n+1)(n+2)$ points being a centre of perspective of two n-ads, and the pole of a corresponding $(n-2)$ fold, in regard to the same quadric.

Ex. 2. Shew that the figure is obtainable by projection of the points of intersection of $(n+2)$ arbitrary $(n-1)$ folds in a space of n dimensions, from an arbitrary point, T, of this. Further, that $(n+2)$ points $P_1 \ldots P_{n+2}$ can be found in this n-fold space, one being arbitrary, such that the $\frac{1}{2}(n+1)(n+2)$ points are of symbols $P_r - P_s (r, s = 1 \ldots n+2)$, where $P_1 \ldots P_{n+2}$ are the symbols of these points.

Ex. 3. Shew that the figure is also obtainable by section, with a $(n-1)$ fold, from that formed by the joins (lines, planes, etc.) of $n+2$ arbitrary points of a space of n dimensions.

Ex. 4. In Ex. 2, if the $(n-1)$ folds be $x_1 = 0, \ldots, x_{n+2} = 0$, the centre of projection, T, being such that $x_1 = x_2 = \ldots = x_{n+2}$, where $\Sigma a_r x_r = 0$, $\Sigma a_r = 0$, shew that there is polarity in regard to the cone $\Sigma a_r x_r^2 = 0$, or $\Sigma a_r u_r^2 = 0$, where $u_1, u_2, \ldots$ are such that any $(n-1)$ fold through T is $\Sigma a_r u_r x_r = 0$. Further, that the point P_m is $(h_1 \ldots h_{m-1} \sigma_m h_{m+1} \ldots h_{n+2})$, where $\Sigma a_r h_r = a_m (h_m - \sigma_m)$, and $h_1 \ldots h_{n+2}$ are arbitrary. Consider the cases $n = 2$, $n = 3$, $n = 4$.

NOTE II

ON THE *HEXAGRAMMUM MYSTICUM* OF PASCAL

If we have in a plane two triads of points, U, V, W and U', V', W', which are in perspective, the joins UU', VV', WW' meeting in a point, and the intersections of the pairs of lines, VW and $V'W'$, WU and $W'U'$, UV and $U'V'$ being in line, then the remaining intersections of the lines VW, WU, UV with the lines $V'W', W'U'$, $U'V'$ are six points which lie on a conic, as we know from what we have in the text called the converse of Pascal's theorem ; and, if we take these six points in any order, the triad formed by one set of alternate joins of these six points is in perspective with the triad formed by the other set of alternate joins, as we know from Pascal's theorem. The axes of perspective of these various pairs of triads are called Pascal lines ; they are sixty in number. They cointersect in sets of three (or of four) in various ways, and these points of intersection lie in sets upon lines, as has gradually appeared from the work of many mathematicians, Steiner, Kirkman, Cayley, Salmon, Veronese, Cremona, and others. The proof of the properties in the plane, though interesting, is intricate. It depends upon successive applications of Desargues' theorem, which we have learnt to regard as deduced from Propositions of Incidence in three dimensions ; and it is the fact that all the results which have elicited attention are such as may be obtained by projection from a figure existing in such a space ; this manner of proof is much simpler than the proof in the plane. Applied by Cayley in a particular way (*Coll. Papers*, vi, pp. 129–134 (1868)), it was applied by Cremona to a figure more general than Pascal's (*Memorie d. r. Acc. d. Lincei*, i, 1877, pp. 854–874). Three distinct points of Cremona's treatment of the figure in three dimensions are, (*a*) That it depends upon a configuration of fifteen lines lying in threes in fifteen planes, of which three pass through each line, (*b*) That this configuration is deducible from a figure of six planes, forming, in Cremonas phrase, the *nocciolo* of the whole, (*c*) That when the equations of these planes are introduced, only one of the identical relations connecting these comes into consideration. It follows from the last that we may regard the six loci as being in a space of four dimensions. This procedure, which Richmond has used, leads to a very simple treatment of the whole matter, which is the more interesting because the figure in four dimensions is one which, as we shall see in a subsequent volume, is of fundamental importance. Cremona's figure of fifteen lines lying by threes in fifteen planes is an example of the (3, 3) correspondence which it is possible to set up (in various ways) between any two sets of fifteen things; this had been considered by Sylvester, in connexion with the problem of naming a function of six letters capable only of six values under permutation of the letters (Sylvester, *Coll. Papers*, I, p. 92 (1844); II, p. 265 (1861)); it is pertinent and instructive for the problem in hand to consider this. In the following note we have (1) dealt with this problem of arrangements, (2) shewn how this can be represented by figures in four, three and two dimensions, (3) pointed out the properties of the figures which lead to the classical properties of the Pascal figure, (4) explained the exact figure, relating to a cubic surface with a node, by which Cremona was led to the generalisation, (5) given some examples of the proof in the plane of the properties of Pascal's figure, and of the configurations arising from some selected portions of the figure in three dimensions, and (6) made reference to the more important original authorities. The reader may prefer to read (4) before the other sections.

(1) Consider six elements, which we denote, at present, by 1, 2, 3, 4, 5, 6. These can be arranged in three pairs, for example 12, 34, 56, wherein, in each pair, the order of the elements is indifferent, and the order of the pairs, in each such set of three pairs, is also indifferent. Such a set of three pairs, involving all the elements, is what was called by Sylvester a *syntheme*, each of the pairs being what he called a *duad*. The total number of possible duads is fifteen; this is also the total number of possible synthemes. It is possible to choose a set of five synthemes which contain, in their aggregate, all the fifteen duads. Such a set of synthemes is called a system. The total number of possible systems is six; if these be all taken, each containing five synthemes, every syntheme will occur twice, in different systems; as has been said, each duad occurs once in each system. Such an arrangement was given by Sylvester (see below). If we assign names to the systems, say P, Q, R, P', Q', R', each syntheme, as occurring in two systems, will correspond to two of these six letters. Thus every one of the fifteen pairs which can be formed by two of these six letters, say, a letter-duad, will correspond to three number-duads, chosen from the possible fifteen duads of numbers, forming a syntheme. The converse is also true; any duad of the numbers being taken, the remaining four numbers can be taken in pairs in three ways; thus any duad occurs in three synthemes, and each of these synthemes enters in two of the systems. The duad thus corresponds to three pairs of systems; or any one of the fifteen number-duads corresponds to three letter-duads, chosen from the possible duads of P, Q, R, P', Q', R'. Therefore, as any

fifteen things can be identified by naming them after the pairs which can be formed from six arbitrary symbols, we can have a (3, 3) correspondence between any two sets of fifteen things. As has been said, the importance of this for the present purpose was emphasized by Cremona (*loc. cit.* pp. 854, 866, 870). A further remark which is of use is that two synthemes which have no duad in common occur together in only one system, since two systems have only one syntheme in common. Two such synthemes thus serve to identify a system.

Of the various possible ways of assigning the names $P, Q, ..., R'$ to the systems, one example is given in the following scheme, which can be read either in rows or columns.

	P	Q	R	P'	Q'	R'
P		14.25.36	16.24.35	13.26.45	12.34.56	15.23.46
Q	14.25.36		15.26.34	12.35.46	16.23.45	13.24.56
R	16.24.35	15.26.34		14.23.56	13.25.46	12.36.45
P'	13.26.45	12.35.46	14.23.56		15.24.36	16.25.34
Q'	12.34.56	16.23.45	13.25.46	15.24.36		14.26.35
R'	15.23.46	13.24.56	12.36.45	16.25.34	14.26.35	

With this table, the correspondence of the pairs of systems with the synthemes can be enumerated at once; for instance, (Q, R) is associated with (15.26.34), and (R, P') with (14.23.56). Conversely, the duads of the numbers are each associated with three duads of letters; for example, (14) with $(PQ . RP' . Q'R')$, since (14) occurs in the same syntheme in P and Q, in the same, other, syntheme in R and P', and in the same, still other, syntheme in Q' and R'. This set of three pairs of letters forms a syntheme, and the synthemes of the letters can similarly be arranged in six systems, with the names 1, 2, 3, 4, 5, 6.

We shall also denote the number-synthemes by the letters $a, b, c, d, a', b', c', d', e, l, m, n, p, q, r$, by the rule which is found by identifying the preceding scheme with the scheme subjoined:

	P	Q	R	P'	Q'	R'
P		e	d	a	b	c
Q	e		d'	a'	b'	c'
R	d	d'		l	m	n
P'	a	a'	l		r	q
Q'	b	b'	m	r		p
R'	c	c'	n	q	p	

Perhaps the easiest way to describe these notations is by the diagram given below in (2), p. 225, or, in another form, in the Frontispiece of the volume.

Now let us consider four of the six systems, considered as two pairs, say, for definiteness Q, R' and Q', R. The synthemes common to the two pairs, that is the syntheme common to Q and R', and the syntheme common to Q' and R, will necessarily have a duad in common. Further this aggregate of two letter-duads, Q, R' and Q', R, may be identified by the common number-duad of the two synthemes, taken with one duad from the first syntheme, and one duad from the second syntheme, so chosen as to have one of its numbers the same as one of the numbers of its companion duad from the first syntheme. In the particular case chosen, Q and R' have common the syntheme $13.24.56$, while Q' and R have common the syntheme $13.25.46$. These synthemes have the duad 13 common; we say that the combination $QR' . Q'R$ may be identified by the symbol $13.24.25$, or by $13.42.46$, or by $13.56.52$, or by $13.65.64$. All this is quite easy to see. For first, two synthemes, that have two duads identical, are themselves identical; and second, two synthemes that have no duad in common both occur in the same system, and nowhere else, as we have remarked. The two synthemes chosen, the former common to one pair of systems, the latter common to another pair, must therefore have a duad common. Using $\alpha, \beta, \gamma, \alpha', \beta', \gamma'$ to denote the numbers $1, 2, ..., 6$, in some order, let these synthemes be $\beta'\gamma' . \alpha\beta . \gamma\alpha'$ and $\beta'\gamma' . \alpha\gamma . \beta\alpha'$. From these we can form the symbol $\beta'\gamma' . \alpha\beta . \alpha\gamma$ (as well as three others). Conversely, if this be given, the formation of the two synthemes having $\beta'\gamma'$ in common and containing, respectively, also $\alpha\beta$ and $\alpha\gamma$, is without ambiguity; so that the symbol identifies the two pairs of systems from which it was formed. Such a symbol, or any one of the other three symbols which identify the same pair of systems, we speak of as defining a T-element. The total number of T-elements is thus $\frac{1}{4}.15.4.3$, or forty-five. Now take the six systems in any particular order, say, $PQ'RP'QR'$, where it is to be understood that this is considered equivalent with its reverse order, $R'QP'RQ'P$, and may be named beginning with any one of its letters, as for instance by $RP'QR'PQ'$; thus the total number of orders is $(6!)/2.6$, or sixty. Then take, with each pair of letters in this order, that pair which is not contiguous with it on either side; thus, with QR' take $Q'R$, with RP' take $R'P$, and with $P'Q$ take PQ'. For each of these three sets of two pairs, form the corresponding T-element. It will be found that these can be represented by symbols of the respective forms $\beta'\gamma' . \alpha\beta . \alpha\gamma$, $\gamma'\alpha' . \alpha\beta . \alpha\gamma$, $\alpha'\beta' . \alpha\beta . \alpha\gamma$. For instance, with the order $PQ'RP'QR'$, the two pairs $QR', Q'R$ give a symbol $13.64.65$, the two pairs $RP', R'P$ give a symbol $23.64.65$, and the two pairs

PQ', $P'Q$ give a symbol $12.64.65$. And these three symbols have in common the duads 64.65, which have the number 6 in common. Conversely consider the aggregate of two duads $\alpha\beta . \alpha\gamma$, having the number α in common; we shew that this leads to a particular order of the symbols $P, Q, ..., R'$. This aggregate belongs to only three symbols of T-elements, namely $\beta'\gamma' . \alpha\beta . \alpha\gamma$, $\gamma'\alpha' . \alpha\beta . \alpha\gamma$ and $\alpha'\beta' . \alpha\beta . \alpha\gamma$. Of these the first arises by combining the two synthemes $\beta'\gamma' . \alpha\beta . \gamma\alpha'$, $\beta'\gamma' . \alpha\gamma . \beta\alpha'$, and there is no ambiguity as to these; for the moment let these synthemes be denoted, respectively, by p and p'. The symbol $\gamma'\alpha' . \alpha\beta . \alpha\gamma$ similarly arises from the two synthemes $\gamma'\alpha' . \alpha\beta . \gamma\beta'$, $\gamma'\alpha' . \alpha\gamma . \beta\beta'$, which we denote, respectively, by q and q'. Lastly, the symbol $\alpha'\beta' . \alpha\beta . \alpha\gamma$ arises from the two synthemes $\alpha'\beta' . \alpha\beta . \gamma\gamma'$, $\alpha'\beta' . \alpha\gamma . \beta\gamma'$, which we denote, respectively, by r and r'. We have, however, remarked that two synthemes, which have no duad in common, determine a particular system. Thus the synthemes q and r', namely $\gamma'\alpha' . \alpha\beta . \gamma\beta'$ and $\alpha'\beta' . \alpha\gamma . \beta\gamma'$, determine one of the systems $P, Q, ..., R'$; this system we may, for a moment, denote by (qr'). It is easy to see that the whole of the six systems are thus determined by such pairs of the six synthemes, which we may arrange in the order $pq' . q'r . rp' . p'q . qr' . r'p$, there being no other pairs of these synthemes which have no duad in common. So arranged, every consecutive pair of these systems determines, with the pair which is not contiguous with it on either side, a T-element. For instance, if we take the systems pq', $q'r$, $p'q$, qr', the first pair have common the syntheme q' or $\gamma'\alpha' . \alpha\gamma . \beta\beta'$, the second pair have common the syntheme q or $\gamma'\alpha' . \alpha\beta . \gamma\beta'$, and these together determine the T-element represented by $\gamma'\alpha' . \alpha\beta . \alpha\gamma$.

We see then that the operation which we carry out when, in Pascal's figure, we form the Pascal line for six points $P, Q, ..., R'$ of a conic, taken in a particular order, can be carried out exactly with the symbols, the systems of synthemes taking the places of the points of the conic. The join of any two of the six Pascal points is replaced by one of the fifteen number-synthemes; the intersection of two of these joins is replaced by a T-element, represented by such a symbol as $\beta'\gamma' . \alpha\beta . \alpha\gamma$; and the Pascal line containing three of these intersections is replaced by a symbol, such as $\alpha\beta . \alpha\gamma$, consisting of two duads having a number (α) in common. We have seen that any such symbol leads back to a particular order of the six letters $P, Q, ..., R'$. It is in accordance with this, that the number of possible symbols $\alpha\beta . \alpha\gamma$ is $6 . 10$ or sixty. It is at once seen in the Pascal figure that through the intersection of two opposite sides of the hexagon there pass four Pascal lines, the two pairs QR', $Q'R$, for example, being non-contiguous in each of the four hexagons $PQ'RP'QR'$, $PQ'RP'R'Q$, $PRQ'P'QR'$, $PRQ'P'R'Q$; this corresponds to the fact remarked that the T-element $\beta'\gamma' . \alpha\beta . \alpha\gamma$ is capable also of the representations $\beta'\gamma' . \gamma\alpha' . \alpha\gamma,\ \beta'\gamma' . \alpha\beta . \beta\alpha',\ \beta'\gamma' . \alpha'\gamma . \alpha'\beta$. The total intersections of the fifteen sides in the Pascal hexagon, in number $\frac{1}{2} 15 . 14$ or 105, consist in fact of ten intersections at each of the six vertices $P, Q, ..., R'$, together with an intersection at each of forty-five T-points.

(2) We now consider a geometrical interpretation of the relations we have described. First, in four dimensions, we can set up a figure of fifteen points lying in threes on fifteen lines, of which three pass through each of the points, beginning in various ways. If we take six general points, say F, G, H, R, S, T, of which the symbols will be subject to one relation, which we write in the form $F + G + H + R + S + T = 0$, there will be fifteen points of symbols each the sum of two of $F, G, ..., T$, and, for instance, the points $F + G, H + R, S + T$ will be in line; of such lines there will be fifteen. Three of these will pass through each point; for instance, the lines $(F+G, H+R, S+T), (F+G, H+S, R+T), (F+G, H+T, R+S)$ pass through the point $F + G$. Denoting the six points by the numbers 1, 2, ..., 6, each duad, such as 12, may be supposed to represent one of these points, and each syntheme, such as 12 . 34 . 56, to represent one of these lines. Or, we may take four arbitrary lines of general position, say, a, b, c, d; every three of these will have, in four dimensions, a single transversal; let the transversal of a, b, c be denoted by d', that of b, c, d by a, and so on, the four transversals being a', b', c', d'. It is then easy to shew that the line, say n, joining the points $(a, b'), (a', b)$ intersects the line, say r, joining the points $(c, d'), (c', d)$; so, the line joining the points $(c, a'), (c', a)$, say m, meets the line joining the points $(b, d'), (b', d)$, say q; and likewise the line joining the points $(b, c'), (b', c)$, say l, meets the line joining the points $(a, d'), (a', d)$, say p. And further that these three points of intersection are in a line, say e. Six fundamental points $F, G, ..., T$, from which these statements can be justified, are obtained by regarding the points $(b', c), (c', a), (a', b), (b, c'), (c, a'), (a, b')$ as being, respectively, the points 23, 31, 12, 56, 64, 45. These six points may in fact be taken arbitrarily to obtain such a figure; denoting them, respectively, by A, B, C, A', B', C', the six consecutive joins of the hexagon $BC'AB'CA'$ are the lines a, b', c, a', b, c'; the lines d, d' are then, respectively, the common transversals of a', b', c', and of a, b, c, and the figure can be completed. The fundamental points are then $F = \frac{1}{2}(B + C - A), R = \frac{1}{2}(B' + C' - A')$, etc. The various relations and notations are represented by the diagram annexed; the fundamental importance of the figure justifies this lengthy description. We shall for clearness denote this figure by Ω. It depends on twenty-four constants. We do not now consider the dually corresponding figure in four dimensions. We consider however the figure obtained by projecting the figure Ω into space of

three dimensions, which we denote by S. It can be constructed, as in the diagram annexed, or as in the diagram given in the Frontispiece, by taking four points which lie in a plane, those denoted by 23, 35, 56, 62; then drawing, through 23, the two arbitrary lines b' and c, and, through 56, the two arbitrary lines b and c'; then, from 35, drawing the line d to meet b' and c', and the line a' to meet b and c, and also, from 62, drawing the line d' to meet b and c, and the line a to meet b' and c'. The remaining incidences of the figure then follow necessarily[1], as the reader may easily see. This figure

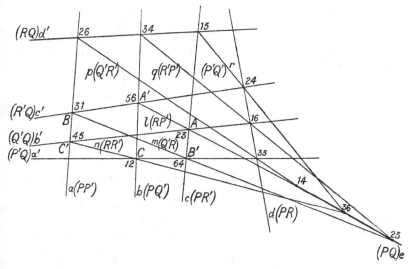

depends on nineteen constants; so that there are ∞^1 possible figures when the six fundamental points, from which it can be constructed, are given. The figure in three dimensions which is the dually corresponding to S will be denoted by S'; it is the figure mainly considered by Cremona (*loc. cit.*), containing fifteen lines lying in threes in fifteen planes, of which three pass through each line. Finally we consider the figure obtained by projecting the figure S' on to a plane, which we denote by ϖ; it is in this figure, ϖ, that the various incidences which arise from Pascal's hexagram are found, and for a figure more general than Pascal's. This plane figure consists of two quadrilaterals, say $ABCD$, $A'B'C'D'$, the intersections of corresponding sides being X of AB and $A'B'$, Y of BC and $B'C'$, Z of CD and $C'D'$ and U of DA and $D'A'$, which are such that the points of intersection, O of XZ and YU, P of AC' and $A'C$ and

---

[1] A proof of this is given, *Proc. Roy. Soc.* LXXXIV, 1911, p. 599.

Q of *BD′* and *B′D*, are in line. This figure depends on fifteen constants; so that there are ∞^3 possible figures when the six fundamental points, from which it can be constructed, are given.

The correspondence between the figures and the Pascal configuration can now be described precisely, as in the preceding section. If we take a particular order of the systems, say *PQ′RP′QR′*, and consider either the figure Ω, or the figure *S*, it is clear that the lines *Q′R*, *QR′*, or *m* and *c′*, are those which join the points 64, 65 to the point 13, the lines *RP′*, *R′P*, or *l* and *c*, are those which join the points 65, 64 to the point 23, and the lines *P′Q*, *PQ′*, or *a′* and *b*, are those which join the points 64, 65 to the point 12. Thus we have three T-planes passing through the line 64 . 65. In the figure *S′*, dually corresponding to *S*, we have correspondingly three T-points lying on a line. Projection of *S′* into the figure ϖ gives then three T-points lying on a line, as in the Pascal configuration. More particularly, each of the systems *P, Q′, ...* consists of five lines, both in the figure *S* and the figure *S′*; the systems *Q, R′* have a definite line in common, as also have the systems *Q′, R*. In order to project *S′* into ϖ, we must take a definite centre of projection, say *O*; from this, a transversal can be drawn to these two lines *QR′* and *Q′R*; it is the intersection of this with the plane of ϖ which gives the T-point. The consecutive lines obtained by the projections of the lines *PQ′, Q′R, RP′, P′Q, QR′, R′P* on to the plane of ϖ, similarly give six lines forming a hexagon; the three points of intersection of the pairs of opposite sides of this lie on a line. The vertices of the hexagon in the plane of ϖ therefore lie on a conic. If we take another order of the systems, for instance, *PQ′RP R′Q*, and project from the same point *O*, we obtain another hexagon, whose vertices lie on a conic, with its Pascal line. This hexagon will not coincide with the former; in the particular instance, we shall have, instead of the vertex obtained by the intersection of the lines which are the projections of *RP′* and *P′Q*, the vertex obtained by the projections of the lines *RP′* and *P′R′*; and so on. As we shew in section (4) below, it is however possible so to specialise the figure as to obtain always the same conic in the plane ϖ.

(3) Dealing now with the figure Ω, or *S*, we have the fifteen points such as 12, which we may call the Cremona points; we also have the fifteen lines, each containing three of these points whose symbols form a syntheme, such as 12 . 34 . 56. Any two of the Cremona points whose symbols have a number in common may be joined by a line; for instance the line joining the points 12 and 13 is such a line. These lines we call the Pascal lines; their number is sixty. A plane containing three Cremona points whose duad symbols are formed with five of the six numbers, as for instance the plane 23 . 64 . 65, is called a T-plane; it contains four Pascal lines (64 . 65;

46.41; 56.51; 14.15), and three such planes pass through any Pascal line. In all there are forty-five such planes. There are however also sets of four Pascal lines which meet in a point. For instance the points 64, 61 are those which, in terms of the six fundamental points F, G, H, R, S, T, have the symbols $T + R, T + F$. The line joining these contains the point $R - F$; this we denote by [14]. Evidently there are fifteen such points, which we call Plücker points; and, for instance, through the point [14], there pass the four Pascal lines 21.24, 31.34, 51.54 and 61.64. The Plücker points lie in threes upon twenty lines; for instance the points [23], [31], [12], whose symbols are, respectively, $G-H, H-F, F-G$, lie upon a line. Such a line is called a g-line, or a Cayley-Salmon line. The Plücker points also lie in sixes upon fifteen planes. For instance the points [63], [64], [65], [45], [53], [34] lie on a plane. Such a plane is called an I-plane, or a Salmon plane. The Pascal lines lie in threes in planes, in two distinct ways. A plane containing three Cremona points whose duad symbols contain only three of the numbers, for instance the plane of the points 23, 31, 12, evidently contains the three Pascal lines 12.13, 23.21, 31.32. Such a plane is called a G-plane, or a Steiner plane; the total number of such planes is twenty. Again a plane containing three Cremona points whose duad symbols all contain one number in common, for instance the plane of the points 41, 42, 43, evidently contains the three Pascal lines 42.43, 43.41, 41.42. Such a plane is called a K-plane, or a Kirkman plane; the total number of such planes is sixty. Lastly, it is necessary to refer to fifteen lines, each joining a Cremona point to the Plücker point whose symbol is formed with the same numbers, for instance the line joining 12 to [12]. Such a line is called an i-line, or a Steiner-Plücker line. The Pascal lines will also be called k-lines. There are certain theorems of incidence among the various elements in addition to those which have been referred to. For instance, a Pascal line, or k-line, lies in one G-plane, and in three K-planes; this fact will be denoted, in the summary we now give, by writing $k...G, 3K$. Conversely a K-plane contains, not only three k-lines, as we have seen, but also one g-line; this fact will be denoted by writing $K...g, 3k$. We may then summarise the definitions, and the incidences referred to, which will be immediately proved, as follows:

> 60 k-lines, Pascal lines, 12, 13,
> 60 K-planes, Kirkman planes, 14, 15, 16;
> 20 g-lines, Cayley-Salmon lines, [12], [13],
> 20 G-planes, Steiner planes, 56, 64, 45;
> 15 i-lines, Steiner-Plücker lines, 12, [12],
> 15 I-planes, Salmon planes, [63], [64], [65].

$$k \ldots G,\, 3K\,; \qquad g \ldots G,\, 3K,\, 3I\,; \qquad i \ldots 4G,$$
$$K \ldots g,\, 3k\,; \qquad G \ldots g,\, 3k,\, 3i\,; \qquad I \ldots 4g.$$

The proof of these last theorems of incidence is immediate:

(1) The k-line 12, 13 lies in the G-plane 23, 31, 12; and in the K-planes (12, 13, 14), (12, 13, 15), (12, 13, 16).

(2) The K-plane 14, 15, 16 contains the g-line ([56], [64], [45]), and the k-lines 15 . 16, 16 . 14, 14 . 15.

(3) The g-line [12], [13] lies in the G-plane (23, 31, 12), it lies in the K-planes (41, 42, 43), (51, 52, 53), (61, 62, 63), and it lies in the I-planes ([12], [13], [14]), ([12], [13], [15]), ([12], [13], [16]).

(4) The G-plane 56, 64, 45 contains the g-line [56], [64], [45], the k-lines 45 . 46, 56 . 54, 64 . 65, and also the i-lines (56, [56]), (64, [64]), (45, [45]).

(5) The i-line (12, [12]) lies in the G-planes (12, 23, 31), (12, 24, 41), (12, 25, 51), (12, 26, 61).

(6) The I-plane ([63], [64], [65]) contains the g-lines ([63], [64], [34]), ([64], [65], [45]), ([65], [63], [35]), ([45], [53], [34]).

It is interesting, however, further, to remark, that the relations expressed by

$$K \ldots g,\, 3k\,; \qquad G \ldots g,\, 3k,\, 3i\,; \qquad I \ldots 4g$$

can be verified by considering only a single threefold space which forms part of Ω, there being in Ω fifteen such spaces; and the relations expressed by

$$k \ldots G,\, 3K\,; \qquad g \ldots G,\, 3K,\, 3I\,; \qquad i \ldots 4G$$

can be verified by considering only elements passing through a single point of Ω, there being fifteen such points, which correspond dually in Ω to the threefold spaces just spoken of. For if we consider the space determined by four of the six fundamental points, say by G, H, S, T, which contains the points 23, 25, 26, 35, 36, 56, [23], [25], [26], [35], [36], [56], it is at once seen that this contains

four K-planes,   four G-planes,   one I-plane;

it contains also

twelve k-lines,   four g-lines,   six i-lines.

Correspondingly, consider elements passing through the Plücker point [14]; there are of such (cf. the Frontispiece of the volume)

four k-lines,   four g-lines,   one i-line;

any plane of the complete figure Ω which contains one of these lines necessarily contains the point [14]. There pass in all through this point

twelve K-planes,   four G-planes,   six I-planes;

one such K-plane is $(21, 24, 23)$; one such G-plane is $(21, 24, 14)$; one such I-plane is $([14], [15], [16])$.

And in fact, if we introduce the equations to the spaces and planes of the figure Ω, we can make the duality suggested by the notation apparent analytically.

If now we suppose that in place of the figure S we consider the figure S', dually corresponding to S in three dimensions, we shall have therein k-lines, g-lines and i-lines, but we shall have K-*points*, instead of K-planes, G-points and I-points. If then we project on to a plane, obtaining the figure ϖ, we shall have therein lines from the lines of the figure S', and points from the points of S'. And therein the six theorems of incidence which we have obtained will continue to hold.

It is this fact which is the outcome of the many investigations made for Pascal's figure, of which the figure ϖ is a generalisation.

(4) We shew now how the exact Pascal's figure can be obtained from a figure in three dimensions; and it will appear at once that this figure is a particular case of the figure S' which we have considered.

Let x, y, z, t be coordinates in three dimensions. Let U be a homogeneous polynomial of the second order in x, y, z only, and M be a homogeneous polynomial of the third order in x, y, z only. Thus $U = 0$ may be regarded as the equation to a conic, in the plane whose equation is $t = 0$; similarly $M = 0$ may be regarded as the equation to a curve in this plane which has the property of being met, by an arbitrary line of this plane, in three points, and is therefore said to be a curve of the third order, or a cubic. This cubic meets the conic in six points, as follows from the elements of the theory of elimination. These points we name, in some order, P, Q', R, P', Q, R'; a cubic curve can be drawn through nine arbitrary points, so that, by proper choice of M, we may regard these six points as arbitrary points of the conic $U = 0$. Now consider the equation $M - tU = 0$, which is homogeneous in the four coordinates x, y, z, t. It represents a surface which is met by an arbitrary line of the threefold space in three points, say, a cubic surface. But, in particular, any line of the threefold space which is drawn through the point of coordinates $(0, 0, 0, 1)$, say, the point D, meets the surface in two coincident points at D, and in a further point; for this reason the point D is called a node of the cubic surface. More particularly, however, there are lines which lie entirely upon the cubic surface: for instance, whatever θ may be, if $(x, y, z, 0)$ be one of the six points for which $U = 0$ and $M = 0$, the point (x, y, z, θ) lies on the cubic surface, since its coordinates satisfy $M - \theta U = 0$, and, for different values of θ, this point is any point of the line joining D to $(x, y, z, 0)$. Let the six lines so obtained, joining D to the six points $P, Q', \ldots$, be named, respectively, p, q', r, p', q, r'; it is easy to see that these are the only lines passing through D which lie on the cubic surface. Then, further, it is at once clear that any plane meets the cubic surface in a cubic curve; thus a plane containing two of the lines $p, q', \ldots, r'$, will contain a further line lying entirely on the cubic surface. Thereby we obtain fifteen further lines of the surface, any one of which may be denoted by a duad symbol, such as qr'. There are however no other lines lying on the surface. For we have enumerated all those which pass through D; suppose then one which does not pass through D, and consider the curve in which the cubic surface is met by the plane joining D to this line; this consists of the line, and a curve of the second order; but this curve of the second order has the property of being met in two points coinciding at D by every line in the plane drawn through D, and must therefore itself consist of two lines intersecting at D. The plane is therefore one of those before considered, drawn through two of the lines $p, q', \ldots, r'$. Now consider the three lines of the surface, q, r' and qr', which lie in a plane. The first line, q, is evidently met by the four lines qp, qr, qp', qq', in addition to qr'; the other line, r', is evidently met by the four lines $r'p, r'r, r'p', r'q'$, in addition to $r'q$. Beside the four lines which meet both q and r' at the point D, there remain then, of the total twenty-one lines of the surface, just six, namely those whose symbols are the duads from p, r, p', q'. These six lines do not lie in the plane of q, r' and qr'; each must meet, then, either q, or r', or qr'; but each meets two lines, other than q and r', of the lines of the surface which pass through D; it cannot, then, for example, meet also the line q, since else three of the lines through D would lie in a plane, and the conic $U = 0$ would break into two lines, which we suppose not to be the case. Thus these six lines all meet the line qr'.

Consider, for instance, rp'; the plane containing qr' and rp' must then meet the cubic surface in another line, by an argument applied above; this other line, meeting both qr' and rp', will have for symbol a duad not containing either q or r', or r or p'; this symbol is, then, pq', the symbols of the three lines qr', rp', pq' containing all the six symbols p, q', r, p', q, r'. In this way we see that there are, beside the plane containing q and r', three planes through the line qr' each containing two other lines with duad symbols. Whence we see that the fifteen lines of the cubic surface, other than those through D, lie in threes in fifteen planes, of which three planes pass through every one of these lines.

If desired, the equations of the fifteen lines can be obtained, when the six points of intersection of the conic $U = 0$, and the cubic $M = 0$ are given. For if $ax + by + cz = 0$, with $t = 0$, be the equation of the join of two of these six points, and we suppose, for instance, that c is not zero, as the plane joining D to this line contains a line of the cubic surface, the substitution for z, in the quotient M/U, of $-(ax+by)/c$, must reduce this to a form $lx+my$; the line in question will then be given by the two equations $ax+by+cz=0$, $t-lx-my=0$.

Having obtained the figure of fifteen planes, and the lines lying in threes of these, we can denote any one of the planes by a duad formed from two of the six numbers 1, 2, 3, 4, 5, 6, and the line of intersection of three of these planes by a syntheme of three duads, such as $12.34.56$, formed from the duads which represent the three planes passing through the line. We shall then have a figure which, combinatorially, has the same properties as the figure S' considered above, specialised however by the fact that the fifteen lines are arranged in sets of five all of which have a common transversal passing through D, any of the lines meeting two of three transversals. These sets of five are the systems of synthemes considered above in (1). But the incidence theorems obtained above will hold for the specialised figure; and, after projection on to the plane $t=0$, these give the properties of the Pascal figure which are the occasion of the present note.

It may be remarked that on a general cubic surface, which does not possess a node, there are sets of fifteen lines such as those here denoted by the duad symbols qr', together with twelve others, each of the six lines here found passing through D being replaced by a pair of skew lines. Moreover, we shall find that the general figure Ω in four dimensions, here considered, is of fundamental importance for the theory of the general cubic surface.

(5) We now give some particular examples of the theory.

Ex. 1. *The pairs of conjugate Steiner planes.* Let $\alpha, \beta, \gamma, \alpha', \beta', \gamma'$ denote the numbers 1, 2, ..., 6, in any order. In the figure Ω, there pass, as we have seen, through each of the three Cremona points, $\beta'\gamma', \gamma'\alpha', \alpha'\beta'$, three of the fifteen lines which are fundamental in the figure (beside two k-lines joining this point to the other two); and the nine lines so obtained meet in threes in the points $\beta\gamma, \gamma\alpha, \alpha\beta$. The two Steiner planes $(\beta'\gamma', \gamma'\alpha', \alpha'\beta')$ and $(\beta\gamma, \gamma\alpha, \alpha\beta)$ are called conjugate. In the Pascal figure there are two corresponding Steiner points, which are in fact conjugate to one another in regard to the fundamental conic. In the figure S', we have two Steiner points; through each of these there pass three Cremona planes, intersecting in pairs in three Pascal lines which pass through the point. The three planes through one of the two Steiner points meet the three planes through the other Steiner point in three triads of lines, of which any two are then in perspective, the axis of perspective being a Pascal line. By projection to the plane ϖ we obtain then a similar result, of which the elementary proof is indicated in more detail below. (Ex. 5. See also v. Staudt, *Crelle*, LXII, p. 142.)

Ex. 2. The Steiner planes and Steiner-Plücker lines. The plane determined by the Cremona points 23, 31, 12, is the same as that determined by the three points F, G, H, of the six fundamental points. The Steiner-Plücker line determined by joining the Cremona point 12 to the Plücker point [12], is that joining the two fundamental points F, G. The twenty Steiner planes are thus all the planes each containing three of the six fundamental points, and the fifteen Steiner-Plücker lines are all the lines each joining two of these points. In the figure S', the Steiner points are then the intersections in threes, and the Steiner-Plücker lines the intersections in twos, of six planes, which, so far as the configuration is concerned, may be taken arbitrarily. The corresponding points and lines in the Pascal figure are then the projection of this very simple configuration. (Cf. p. 218.)

Ex. 3. The separation of the complete figure into six figures. In the figure Ω, the notation at once suggests that we consider together the set of five Cremona points such as 12, 13, 14, 15, 16; and there will be six such sets, each point belonging to two sets. The five points of a set are such that the join of any two is a Pascal line, and the plane of any three is a Kirkman plane. In the figure S', we have then six sets of five Cremona planes, of which any two meet in a Pascal line, and any three in a Kirkman point. In the plane ϖ, the sixty Pascal lines are thus divided into six sets of ten, meeting in threes in ten Kirkman points, each of the Pascal lines containing three of the Kirkman points. The configuration, in each of the six partial figures is that of two triads in perspective, with their centre and axis of perspective. These partial figures were considered by Veronese.

Ex. 4. Tetrads of Steiner points each in threefold perspective with a tetrad of Kirkman points. In the figure Ω, the six Cremona points 23, 31, 12, 41, 42, 43, which lie in the threefold space of the four fundamental points F, G, H, R, give four Steiner planes,

$\alpha = (24, 43, 32)$, $\beta = (34, 41, 13)$, $\gamma = (14, 42, 21)$, $\delta = (23, 31, 12)$, and four Kirkman planes,

$\alpha' = (12, 13, 14)$, $\beta' = (23, 21, 24)$, $\gamma' = (31, 32, 34)$, $\delta' = (41, 42, 43)$. The planes α, α' meet in the *g*-line ([23], [24], [34]), the planes β, β' meet in the *g*-line ([31], [34], [14]), the planes γ, γ' meet in the *g*-line ([12], [14], [24]), and the planes δ, δ' meet in the *g*-line ([12], [13], [23]); and these are the four *g*-lines which lie in the *I*-plane ([41], [42], [43]).

Hence if we consider the figure S', we have two tetrads of points, one formed by Steiner points, and the other by Kirkman points, which are in perspective, the lines joining corresponding points of the two tetrads being *g*-lines, and the centre of perspective being an *I*-point.

But in fact, in this figure S', the two tetrads are in perspective in three other ways. In this figure, α, β, ..., α', β', ... are points; the lines (α, δ'), (β, γ'), (γ, β'), (δ, α') also meet in a point, namely the point $(12, 13, [14])$, or $(12, 13, 56)$, a T-point. This fact we may denote by writing

$$12, 13, [14] \begin{cases} 24, 43, 32 & 34, 14, 31 & 14, 24, 12 & 23, 31, 12 \\ 41, 42, 43 & 31, 32, 34 & 23, 21, 24 & 12, 13, 14 \end{cases}$$

The joins of corresponding points of these two tetrads are, respectively, the Pascal lines 42.43, 31.34, 21.24, 12.13.

Similarly the tetrads $(\alpha, \beta, \gamma, \delta)$, $(\gamma', \delta', \alpha', \beta')$ are in perspective from the point $(23, 21, [24])$, or $(23, 21, 56)$, and the tetrads $(\alpha, \beta, \gamma, \delta)$, $(\beta', \alpha', \delta', \gamma')$ are in perspective from $(31, 32, [34])$, or $(31, 32, 56)$. The two given tetrads, and that formed by the four centres of perspective, form, therefore, what is known as a desmic system of three tetrads, of which the points can be supposed to have symbols of the form (cf. p. 213)

$$(P, Q, R, S), (-P+Q+R+S, P-Q+R+S, P+Q-R+S, P+Q+R-S),$$
$$(P+Q+R+S, P-Q-R+S, -P+Q-R+S, -P-Q+R+S).$$

Ex. 5. Relation of the Cayley-Salmon lines and the Salmon planes.
The fifteen Plücker points in the figure Ω lie in threes on the twenty Cayley-Salmon lines, *g*-lines, and in threes also on the fifteen Salmon planes, or *I*-planes, forming a figure such as that occurring in the proof of Desargues' theorem for two triads in one plane (cf. p. 214).

The figure is evidently dual with itself, in threefold space, and has the same description in S' and in ϖ. Through each of the fifteen points there pass, in S or S', six planes and four lines, each of the twenty lines contains three points and lies in three planes, each plane contains four of the lines and six of the points. The figure has been considered in Note I (§ 3).

Ex. 6. Let U, V, W and U', V', W' be two triads of points in a plane, the lines UU', VV', WW' meeting in G; let the points $(UV, W'U')$, $(VW, U'V')$, $(WU, V'W')$ be denoted, respectively, by P, Q, R, and the points $(U'V', WU)$, $(V'W', UV)$, $(W'U', VW)$ be denoted, respectively, by P', Q', R'; so that, by the converse of Pascal's theorem, the six points P, Q', R, P', Q, R' lie on a conic. Prove, (1), that QQ', RR', UU' meet in a point, say, L; and, similarly, RR', PP', VV' meet in a point, say, M; and, likewise, PP', QQ', WW' meet in a point, say, N. Let the points $(PQ, R'P'), (QR, P'Q), (RP, Q'R')$ be denoted, respectively, by X, Y, Z; and the points $(P'Q', RP), (QR', PQ), (RP', QR)$ be denoted, respectively, by X', Y', Z'. Prove, (2), that X, L, X' are in line, as also Y, M, Y' and Z, N, Z'; and that

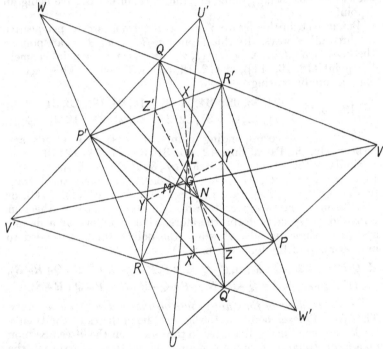

the lines XLX', YMY', ZNZ' meet in a point. Prove also, (3), that the lines YZ, PP', QR' meet in a point, as do the lines $Y'Z', PP', Q'R$; that the lines ZX, QQ', RP' meet in a point, as do $Z'X', QQ', R'P$; and that the lines XY, RR', PQ' meet in a point, as do the lines $X'Y', RR', P'Q$. Thus any two of the triads $U, V, W; L, M, N; X, Y, Z$ are in perspective, as also are any two of the triads $U', V', W'; L, M, N; X', Y', Z'$, the lines XU, YV, ZW meeting in a point, as do $X'U, Y'V', Z'W'$.

Shew also, (4), that the lines UU', VV', WW' are Pascal lines each for a proper order of the six points P, Q', etc., on the conic, their point of meeting, G, being a Steiner point; and, (5), that the axis of perspective of the triads U, V, W and U', V', W' is a Pascal line;

that the axis of perspective of the triads U, V, W and L, M, N is a Pascal line, and that the axis of perspective of the triads U', V', W' and L, M, N is a Pascal line; and that these three Pascal lines meet in a Steiner point. Shew also, (6), that the lines XLX', YMY', ZNZ' are Pascal lines, meeting in a Kirkman point, that the lines XU, YV, ZW are Pascal lines, meeting in a Kirkman point, and that the lines $X'U'$, $Y'V'$, $Z'W'$ are Pascal lines, meeting in a Kirkman point.

These facts may all be obtained by application of Desargues' theorem.

They may also be obtained by shewing that, with that association of P, Q, R, etc. with the systems which has been adopted in this note, the points U, V, W, U', V', W' are the respective T-points $(56, 21, 23)$, $(56, 12, 13)$, $(56, 31, 32)$, $(64, 21, 23)$, $(64, 12, 13)$, $(64, 31, 32)$, so that the lines UU', VV', WW' are the respective Pascal lines $(23, 21)$, $(12, 13)$, $(31, 32)$. Then that the points L, M, N are the respective T-points $(45, 23, 21)$, $(45, 12, 13)$, $(45, 13, 23)$. Then that the points X, Y, Z are the respective T-points $(25, 63, 61)$, $(15, 62, 63)$, $(35, 61, 62)$, and the points X', Y', Z' are the respective T-points $(24, 61, 63)$, $(14, 62, 63)$, $(34, 61, 62)$, so that the lines XX', YY', ZZ' are, respectively, the Pascal lines $(63, 61)$, $(62, 63)$, $(61, 62)$ and the lines YZ, $Y'Z'$ are, respectively, the Pascal lines $(24, 26)$, $(25, 26)$, with similar results for ZX, $Z'X'$, XY, $X'Y'$. The points (QR', PP'), (RP', QQ'), (PQ', RR') are, respectively, the T-points $(31, 56, 54)$, $(23, 56, 54)$, $(12, 56, 54)$; the lines XU, $X'U'$, YV are, respectively, the Pascal lines $(41, 43)$, $(51, 53)$, $(42, 43)$, and so on.

(6) For the literature of the matter, the reader may consult:

Pascal, *Essai pour les coniques* (1640), *Oeuvres*, Brunschvigg et Boutroux, Paris, 1908, I, p. 245.

Brianchon, *Mémoire sur les lignes du 2ᵉ ordre* (1806).

Steiner, *Gergonne's Ann.* XVIII (1828), *Ges. Werke*, I, p. 451 (1832).

Plücker, *Ueber ein neues Princip der Geometrie*, *Crelle*, V (1830), p. 274.

Hesse, *Crelle*, XXIV (1842), p. 40; *Crelle*, XLI (1851), p. 269; *Crelle*, LXVIII (1868), p. 193, and *Crelle*, LXXV (1873), p. 1.

Cayley, various notes, beginning 1846 (*Coll. Papers*, I, pp. 322, 356); see particularly *Papers*, VI (1868), p. 129.

Grossman, *Crelle*, LVIII (1861), p. 174.

Von Staudt, *Crelle*, LXII (1863), p. 142.

Bauer, *Abh. der k. bayer. Ak. d. Wiss.* (Munich) XI (1874), p. 111.

Kirkman, *Manchester Courier*, June and August, 1849; *Camb. and Dub. Math. Jour.* V (1850), p. 185.

Veronese, *Memorie d. r. Acc. d. Lincei*, I (1877), pp. 649–703; and Cremona, in the same volume, pp. 854–874.

Sylvester, *Coll. Papers*, I, p. 92, and II, p. 265.

Salmon, *Conic Sections*, 1879, pp. 379–383.

Klug, L., *Die Configuration des Pascal'schen Sechseckes*, Kolozsvár (A. A. Ajtai), 1898, pp. 1–132.

Cremona, *Math. Annal.* XIII (1878), p. 301.

Castelnuovo, *Atti Ist. Venet.* v (1887), p. 1249, and VI (1888), p. 525.

Caporali, *Memorie di geometria*, Naples, 1888, pp. 135, 236, 252.

Richmond, *Trans. Camb. Phil. Soc.* XV (1894), pp. 267–302 ; also *Quart. J. of Math.* XXXI (1899), pp. 125–160, and *Math. Annal.* LIII (1900), pp. 161–176. Also *Quart. J. of Math.* XXXIV (1903), pp. 117–154.

Acknowledgments are due to Mr F. P. White, St John's College, Cambridge, for assistance in the preparation of this note.

The general properties of the four-dimensional figure will concern us subsequently. We may however refer to

Segre, *Rend. Acc. Lincei*, III (1887), pp. 149–153, and *Atti Acc. sc. Torino*, XXII, 1887, pp. 791–801.

Schoute, *Amsterdam Proc. Sci. K. Akad. Wet.* IV (1902), pp. 203–214; 251–264.

NOTE III

IN REGARD TO THE LITERATURE FOR NON-EUCLIDEAN GEOMETRY

The following references may be of use, as indication of ways in which the introduction given in Chapter v may be amplified:

R. Bonola, *Non-Euclidean geometry, a critical and historical study*, English by H. S. Carslaw (Open Court Publishing Co., Chicago, 1912, pp. 1–268).

R. Bonola, *Sulla teoria delle parallele e sulle geometrie non-euclidee*, in Vol. i, pp. 248–362, of Enriques' *Questioni riguardanti le matematiche elementari*, 1912.

D. M. Y. Sommerville, *The elements of non-Euclidean geometry*, 1914, pp.1–274; and, by the same author, *Bibliography of non-Euclidean geometry*, St Andrews, 1911, pp. 1–403.

Simon Newcomb, *Elementary theorems relating to...space...of uniform...curvature in the fourth dimension*, Crelle's Journal, lxxxiii, 1877, pp. 293–299.

H. Helmholtz, *Ueber die Thatsachen die der Geometrie zu Grunde liegen*, 1868, Ges. Wiss. Abhand., Band ii.

B. Riemann, *Ueber die Hypothesen welche der Geometrie zu Grunde liegen*, 1854, Ges. Werke.

F. Enriques, *Prinzipien der Geometrie*, Enzykl. der Math. Wiss. iii. 1. 1.

F. Klein, *Nichteuklidische Geometrie*, Göttingen, 1913 ; *Elementarmathematik*, 1913 ; many papers in the *Math. Annal.*; and forthcoming *Mathematical Works*.

D. Hilbert, *Grundlagen der Geometrie* (First Edit. 1899).

F. Schur, *Grundlagen der Geometrie*, 1909.

N. I. Lobatschewsky, *Geometrische Untersuchungen zur Theorie der Parallellinien*, 1840 (facsimile, Mayer u. Müller, 1887).

    Pangéométrie ou précis de géométrie fondée sur une théorie générale et rigoureuse des parallèles, 1855 (reprint, Ostwald's Klassiker, 1902).

    Geometrical works of Lobatschewsky (Kasan, 1883–1886).

F. Engel, *N. I. Lobatschewsky, zwei geometrische Abhandlungen..., mit einer Biographie*, 1898, pp. 1–476.

W. Bolyai, *Tentamen juventutem studiosam in elementa matheseos purae* (1832), with appendix by J. Bolyai, *Appendix scientiam spatii absolute veram exhibens*.

J. Frischauf, *Absolute Geometrie nach Johann Bolyai* (1872).

P. Stäckel, *W. u. J. Bolyai, geometrische Untersuchungen*, 1913, two parts, each pp. 1–281 (see also Stäckel, *Math. u. Naturw. Ber. aus Ungarn*, xvii, 1901, and xviii, 1902. Also Stäckel u. Engel, *Math. Annal.* xlix, 1897, p. 149).

F. Engel u. P. Stäckel, *Die Theorie der Parallellinien von Euclid bis auf Gauss*, 1895, pp. 1–325.

C. A. F. Peters, *Briefwechsel zwischen K. F. Gauss u. H. C. Schumacher* (6 vols.) 1860–1865). See Gauss, *Ges. Werke*.

In what is said in the text, references to non-Euclidean geometry in three dimensions are excluded, as arising subsequently. Also questions of Transformation and Group-Theory are not considered (see S. Lie, *Transformationsgruppen*, iii, 1893, which gives an exhaustive analytical study of the Riemann-Helmholtz problem). Further, the arbitrary limitation of the symbols adopted in Chapters iv and v leads to the omission of many general logical questions. Cf., for example, Dehn, *Math. Annal.* liii (1900), pp. 404–439, and *Math. Annal.* lx (1905), pp. 166–174 (and Hilbert's *Grundlagen*).

NOTE IV

REMARKS AND CORRECTIONS OF VOLUME I

p. 10. Cf. p. 218 of the present Volume; and Veronese, *Math. Annal.* xix (1882), p. 161

p. 12, line 3 from the bottom. For CPR read CPQ.

p. 20, line 16 f. b. Cf. Zeuthen, *Compt. rend.* cxxv (1897), pp. 639, 858. This reference I owe to the kindness of Professor F. Schur.

p. 26, in the diagram, the upper M should be N.

p. 30, line 12. For G, D_1, A_1 read C_1, D_1, A_1.

p. 31, last line. For L, K_1, Q_1 read L_1, K_1, Q_1.

p. 38. Cf. p. 218 of the present Volume.

p. 41, here, and especially on p. 149, reference should have been made to Grassman, *Crelle*, xlix, § 4, 1855, p. 55, who shews how, by several projections in succession, to pass from one plane to another so that four given points of one become four given points of the other. Given two sets of $(n+2)$ points in two n-fold spaces, we can pass from one to the other by a succession of $(n+1)$ perspectivities, by each of which $(n+2)$ points in one n-fold space become $(n+2)$ points in another such space. A sufficient representation of a general method, easily stated in geometrical terms, will be obtained by considering A, B, C, D, O, in one threefold space, and A', B', C', D', O' in another, having, by one perspectivity, arranged that these have one point, O, or O', in common. Taking as successive centres of perspectivity the points of symbols $Z = D - D'$, $Y = C - C'$, $X = B - B'$, we evidently have the results expressed by

$$Z\,(A', \qquad\qquad B', C', D', O) = (-D+D'+A', \qquad B', C', D, O),$$
$$Y\,(-D+D'+A', \qquad B', C', D', O) = (-C+C'-D+D'+A', B', C, D, O),$$
$$X\,(-C+C'-D+D'+A', B', C, D, O) = (A, \qquad\qquad B, C, D, O),$$

where we take
$$O = A'+B'+C'+D' = (-D+D'+A')+B'+C'+D$$
$$= (-C+C'-D+D'+A')+B'+C+D$$
$$= A+B+C+D.$$

p. 52, line 11. For PYY_1 read QYY_1.

p. 76, line 10. For $O+bB$ read $O+bU$.

p. 102, line 14. For AC read AB.

p. 118, line 8 f. b. The reader may consult König, *Arch. Math. Phys.* xix, 1912, p. 214; also Veblen, *Analysis Situs*, Amer. Math. Soc. publication, 1922.

p. 120, lines 9 and 10 f. b. For "these being continued indefinitely beyond P and Q, respectively," read "the former of these being continued beyond P." See the footnote inserted at p. 186, and p. 185, of the present Volume.

p. 122, line 3 f. b. For "do satisfy" read "do not satisfy."

p. 131, line 8 f. b. For $ADD'B$ read $ABDD'$.

p. 149. See what is said above under p. 41.

p. 151, lines 20 and 21. $A\,(B, C, D_1, P)$ and $A'\,(B', C', D_1', P')$ should be interchanged, as also A and A' in line 21.

p. 154, line 23. For $\sigma = \lambda$ read $\sigma = \xi$.

p. 159, lines 13, 14. Omit "of these."

p. 166, line 2. For $(B'\,B, C, A)$ read (B', B, C, A).

p. 178, line 15. For "tursus" read "rursus."

p. 180. Under (f), the references to Cayley, *Coll. Papers*, Vol. i, should have been to pp. 55, 317, 356. At the time of writing I was not aware of the publication (Jan. 1921) of the important Article by C. Segre, in the *Enzykl. Math. Wiss.* (iii, 2, 7), *Mehrdimensionale Räume*, pp. 769–972.

p. 182. Under pp. 165–175, reference to Lüroth, *Math. Annal.* xiii, 1878, p. 303, should have been included.

p. 183. Under Pappus' theorem, add 87–93.

INDEX

(The numbers refer to pages. See also pp. 237, 238.)

CAMBRIDGE: PRINTED BY
J. B. PEACE, M.A.,
AT THE UNIVERSITY PRESS

Printed in the United States
By Bookmasters